Groundwater - New Advances and Challenges

Edited by Jamila Tarhouni

Published in London, United Kingdom

Groundwater - New Advances and Challenges
http://dx.doi.org/10.5772/intechopen.104054
Edited by Jamila Tarhouni

Contributors
Shobha Kumari Yadav, Robert Merton Stwalley III, Tyler J. McPheron, Grace L. Baldwin Kan-Uge, Zackariah J. Horn, Adewale Allen Sokan-Adeaga, Titus Oladapo Okareh, Tosin Akin-Brandom, Micheal Ayodeji Sokan-Adeaga, Eniola Deborah Sokan-Adeaga, Luis Manuel Ferreira-Gomes, Luís José Andrade Pais, Pedro Jorge Coelho Ferreira, Alcino Sousa Oliveira, John Fernando Escobar, Darío Naranjo-Fernández, Sergio Lopera, Orlando Mauricio Quiroz, Alonso Ocampo, German Zarate

First published in London, United Kingdom, 2023 by IntechOpen
IntechOpen is the global imprint of INTECHOPEN LIMITED, registered in England and Wales, registration number: 11086078, 5 Princes Gate Court, London, SW7 2QJ, United Kingdom

British Library Cataloguing-in-Publication Data
A catalogue record for this book is available from the British Library

Additional hard and PDF copies can be obtained from orders@intechopen.com

Groundwater - New Advances and Challenges
Edited by Jamila Tarhouni
p. cm.
Print ISBN 978-1-80355-483-9
Online ISBN 978-1-80355-484-6
eBook (PDF) ISBN 978-1-80355-485-3

For EU product safety concerns:
IN TECH d.o.o., Prolaz Marije Krucifikse Kozulić 3, 51000 Rijeka, Croatia,
info@intechopen.com or visit our website at intechopen.com.

Meet the editor

Prof. Tarhouni received her DEA in Hydrodynamics, Hydrology, and Geophysics from the Scientific and Medical University of Grenoble, France, and a Ph.D. in Water and Land Management from the University of Ghent, Belgium. She is a Professor of Groundwater Modeling and Hydraulics and Director of Sciences and Technologies for the Water Laboratory at the National Agronomic Institute of Tunisia.

Contents

Preface **IX**

Chapter 1 **1**
Land Cover Change and Its Impact on Groundwater Resources: Findings and Recommendations
by Shobha Kumari Yadav

Chapter 2 **31**
Assessment of Heavy Metals Contamination in Groundwater and Its Implications for Public Health Education: A Case Study of an Industrial Area in Southwestern Nigeria
by Titus Oladapo Okareh, Adewale Allen Sokan-Adeaga, Tosin Akin-Brandom, Micheal Ayodeji Sokan-Adeaga and Eniola Deborah Sokan-Adeaga

Chapter 3 **67**
Cost-Effective, Sanitary Shallow Water Wells for Agriculture and Small Communities Using Mechanized Tube Well Installation
by Grace L. Baldwin Kan-Uge, Tyler J. McPheron, Zackariah J. Horn and Robert M. Stwalley III

Chapter 4 **101**
The Importance of Salty Groundwater in the Supply of Thalassotherapy: The Case of Portugal
by Luís Manuel Ferreira-Gomes, Luís José Andrade Pais and Pedro Jorge Coelho Ferreira

Chapter 5 **129**
Stability over Time of the Quality of the Sulfurous Groundwater from the Deep Aquifer System That Supplies the Longroiva Medical Spa (Portugal)
by Pedro Jorge Coelho Ferreira, Luis Manuel Ferreira-Gomes and Alcino Sousa Oliveira

Chapter 6 **159**
Aquifer Management in Hydrocarbon Exploitation Operations
by John Fernando Escobar, Darío Naranjo-Fernández, Sergio Lopera, Orlando Mauricio Quiroz, Alonso Ocampo and German Zarate

Preface

The groundwater potential of outcropping and deep geological layers is very important for socioeconomic development, especially in semi-arid and arid regions. Improving knowledge of this potential and the risks related to entropic activities as well as climate change is a challenge due to the tremendous number of factors affecting the behavior of the aquifers.

Evaluation of groundwater potential has, in fact, made great progress following the development of prospecting technologies as well as the elaboration of advanced methods for interpretation and 3D representation. The very heterogeneous properties of the aquifers have led to the need to implement suitable prospecting methods and specific analyses to assess the capacity of these formations to supply water. The objectives are to improve knowledge of the structure and hydrogeological functioning of the aquifers that constitute the basis of groundwater flow through complex porous media structures. Indeed, advanced groundwater tools are applied either for a better understanding of hydrogeological systems or for the prediction of their future behavior and, therefore, for the assessment of sustainable exploitation of groundwater. Newly available tools apply complementary methods that allow integrating, manipulating, and interpreting existing or newly collected data to overcome limitations related to large datasets.

Consequently to climate change, land use and land cover may be negatively impacted and may induce large disturbances in the hydrological cycle. More specifically, groundwater recharge estimation under land cover and climate change is becoming a major hazard. Advanced informatics tools such as geographic information systems (GIS) allow the management of land cover and land use changes. As a result, sustainable use of groundwater resources as well as their adaptation and mitigation to climate change become possible.

Groundwater is exposed to other risks than those related to climate change, in particular the risks of pollution. The protection of groundwater against the risks of pollution is a necessity to ensure access to water safe for drinking and household use. To achieve this, the challenge lies both in a better characterization of the risks as well as in healthy groundwater exploitation conditions.

This book develops and applies various approaches and tools to study and investigate the impacts of climate change and pollution on groundwater. It also presents sustainable techniques for accessing and managing groundwater resources.

Chapter 1 reviews the status, prospects, and challenges of land cover change and its impacts on groundwater recharge, water quality and quantity, and groundwater storage.

Chapter 2 presents an example of groundwater quality assessment in general and an evaluation of heavy metal toxicity risks, more specifically those induced by the industrial hub of Shagamu, Ogun State, in southwestern Nigeria. It also examines the health implications of heavy metal toxicity on the local residents.

Chapter 3 outlines water access problems from a global perspective, describes the traditional means of construction for sanitary water wells in remote areas and their relative costs, and details recent advancements and potential cost savings provided by a simple mechanized means to install tube wells in shallow water table areas.

Chapter 4 analyzes the importance of salty groundwater for thalassotherapy in Portugal.

Chapter 5 is also dedicated to thalassotherapy in Portugal. It analyzes the stability over time of the quality of the sulphurous groundwater from a deep aquifer using the case example of the Longroiva spa.

Chapter 6 presents adapted aquifer management operations in hydrocarbon exploitation, such as disposals, injection, and enhanced oil recovery. For this book, the proposed water recuperation and recycling process can be found online along with the announced objectives related to sustainable water use and management.

Jamila Tarhouni
Sciences and Technologies for Water Laboratory,
National Agronomic Institute of Tunisia,
Carthage University,
Tunis, Tunisia

Chapter 1

Land Cover Change and Its Impact on Groundwater Resources: Findings and Recommendations

Shobha Kumari Yadav

Abstract

Globally, the climate is becoming drier and wetter because of climate change. Variations in land use and land cover (LULC) brought on by humans have impacted hydrological elements, including recharge and runoff, throughout the past few decades. Agriculture, forestry, urbanization, recreational activities, and industrialization are all land uses that impact groundwater resources. For example, anthropogenic activities have an increased impact on impervious surfaces and storm drains, which divert precipitation away from highways. Similarly, groundwater resources are negatively impacted by the increased urbanization of areas in two fundamental ways: first, by blocking up aquifers with concrete, which prevents natural recharge; second, by polluting groundwater through drainage leaks and industrial waste and effluents. Therefore, the long-term temporal and seasonal variations in LULC change significantly impact groundwater flow dynamics. Numerous factors influence LULC change, including hard-to-follow social and biophysical processes, that ultimately lead to a complex and dynamic system. As a result, an evaluation of the effects of LULC changes on recharge is required to manage groundwater resources to be sustainable.

Keywords: groundwater, land use and land cover, social and biophysical process anthropogenic activities, sustainability, land cover change

1. Introduction

Groundwater is the major water source on Earth [1]. It is an essential source of fresh water for domestic and agricultural usage [2, 3], and it is crucial for the sustainability of the economy and the supply of food [4]. It also plays a crucial role in aquatic ecosystems that have interconnections with surface water [5]. In many parts of the world, the groundwater table is dropping, and the water quality is degrading [6]. Because of excessive groundwater use for agriculture and other unsustainable purposes, groundwater depletion has been rising globally [7, 8] and groundwater contamination is becoming prominent. Groundwater is freshwater that is found in the underground layers of water-bearing porous rock or unconsolidated materials in the aquifer systems [9]. Groundwater formation and flow are influenced by a number of variables, including lithology, topography, geological structures, weathering depth, the size of fractures, slope, drainage, landforms, LULC, elevation, rainfall, and other climatic conditions [10–12].

Approximately 2 billion people throughout the world depend primarily on groundwater for domestic and agricultural needs [13]. As a result, groundwater is crucial for irrigated agriculture and for ensuring the safety of the world's food security. Compared to other economic sectors, agriculture utilizes the most freshwater. It accounts for 90% of freshwater consumption and nearly 70% of the world's freshwater withdrawals [14, 15]. The annual groundwater use for irrigation is 545 km^3 of which 43% of the water used annually comes from groundwater [3]. In many areas, groundwater may be the sole supply of water that is always present. When energy and pumping resources are readily accessible, groundwater is frequently the only source of water. Additionally, it serves as a buffer against short- and long-term fluctuations in surface water availability brought on by climatic variability. The usage of groundwater is influenced by variables, including accessibility, transportability, cost-effectiveness, and availability. The main reasons why people choose to use groundwater water are reliable supplies and reasonable prices [16].

However, groundwater quality is declining due to rising water demand, urbanization, changing land use and land cover, and climate change. Changes in land use and land cover (LULC) are among the most significant anthropogenic interventions LULC reflects the characteristics that are spread both naturally and intentionally on the surface of the Earth, such as vegetation in forests, water bodies, and human structures [17]. Groundwater is affected by LULC changes *via* changes in the composition of the water balance [18, 19]. Agricultural expansion is one of the major LULC change the world has witnessed in the last few decades. Globally, around 5% of the land has been converted to agricultural land [20]. For instance, growing agricultural irrigation in the Texas High Plains of the United States has enhanced output but at the expense of falling water levels, endangering the long-term viability of the Ogallala Aquifer as a major source of water for irrigation [21]. In a single county in the High Plains of Kansas in the United States [22, 23], looked at the link between groundwater depletion and agricultural land use change and concluded that groundwater depletion was caused by land use change. In the southwestern United States [24], distinguished between irrigated and nonirrigated agricultural ecosystems and established many tiers of groundwater recharge rates as a function of LULC change. Further, the area of irrigated cropland has expanded by 460% globally over the previous three centuries [25], while the quantity of groundwater used for agricultural irrigation each year is 3300 km^3 [26] and the amount of evapotranspiration drawdown has decreased by a factor of 2 [25]. The effects of LULC change on groundwater hydrology have previously been studied using both field experiments [24] and hydrologic modeling applications [27, 28]. The benefit of spatially dispersed hydrologic models is that they can take into consideration the geographical patterns of LULC's hydrological impact [29]. Few studies take future land use change into consideration, despite the fact that the hydrological models may offer predictions about the future.

While it is a significant supply of freshwater for household, agricultural, and commercial applications, the effects of LULC change on groundwater recharge are not adequately understood, which leads to groundwater depletion [30]. Land use change is a complex, dynamic process, which has direct impacts on soil, water, and the atmosphere [31]. The most urgent problem of the twenty-first century in terms of groundwater monitoring and accurate projections is the rapidly changing LULC. LULC change is becoming a major ecological concern, particularly the conversion of natural vegetation into croplands or the deterioration of land into barren. Therefore, understanding the impacts of LULC change on the groundwater is needed for the optimal management of natural resources [24]. This chapter review the current status, prospects, and challenge of land cover change and its impact on groundwater resources. The effects of human climate change on groundwater resources are the subject of this review [32], hence they are not discussed in this chapter.

2. Literature review

It is becoming more widely accepted that changes in LULC have an impact on groundwater. Various forms of land-use categories and subcategories that have the potential to impact groundwater resource are presented in **Figure 1**. Understanding how LULC interacts with growing natural and human activities is crucial because it significantly impacts groundwater resources. The worldwide groundwater scenario has been altered by LULC change, with reports of irregular recharging [35], declining groundwater quality [36, 37], and solute transfer in the unsaturated or vadose zone [24]. Particularly, the fundamental nature of the global water cycle has been greatly altered by climate-LULC interaction brought on by human-induced fast LULC change [38]. Further, changes in LULC are known to have an effect on the relationship between groundwater and surface water. Urbanization, overgrazing by animals, subsistence agriculture, commercial agricultural growth, and the removal of forests for firewood are some of the changes that have a significant influence on groundwater resources, land productivity, and ecosystem degradation [39]. However, these effects are poorly understood [24, 40].

Numerous studies have attempted to estimate the possible influence of LULC on groundwater processes. Dams et al. [41] used the CLUE-S (the Conversion of Land Use and its Effects at Small regional scale) model coupled with WetSpass and MODFLOW to demonstrate the effects of land use change on the groundwater system

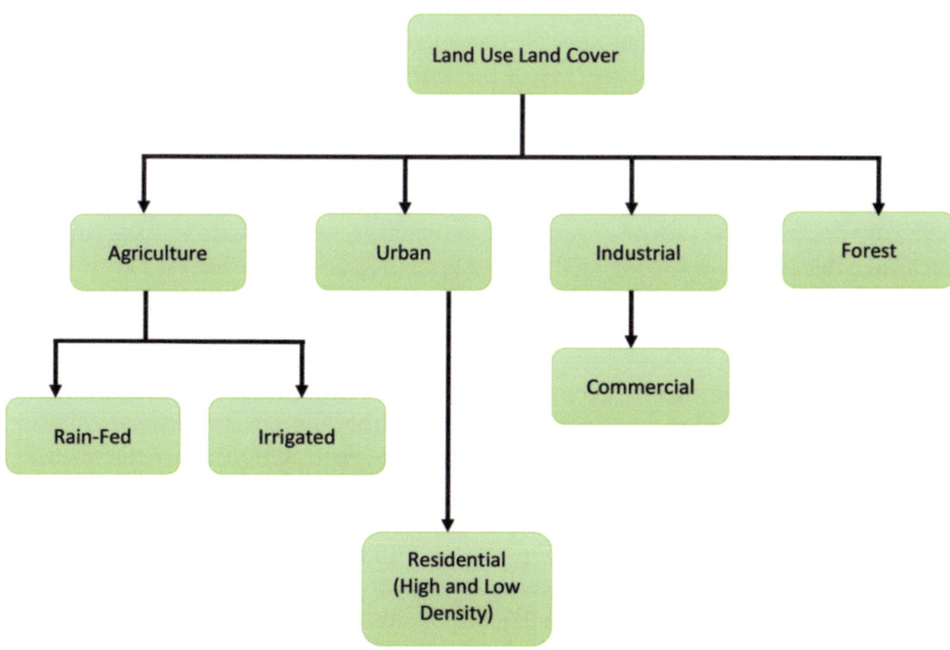

Figure 1.
Major land-use categories that have implications for groundwater resources (adapted from [33, 34]).

of the Kleine Nete watershed, Belgium. Near the largest cities in their research region, they discovered significant alterations in the groundwater. Using the SWAT model Mkaya et al. [42], assessed the influence of land use change on catchment hydrology in Taita Hills, Kenya. Their study showed that the impact of LULC increased surface runoff and sediment output within the watershed. Eckhardt et al. [43] employed the SWAT-G model in the Dill catchment in southeast Germany and demonstrated a 50% drop in mean groundwater recharge and streamflow.

Groundwater recharge rates and mechanisms are significantly altered when land is converted for agriculture, whether it be irrigated or rain-fed [25, 44]. Many parts of the world have experienced substantial changes in water balances, including groundwater recharge, as a result of the removal of native vegetation and establishment of crops, which often have shorter root systems than the plants they replace [45, 46]. This often leads to increases in groundwater recharge rates of one to two orders of magnitude in dry and semiarid regions [25, 47, 48]. Significant negative effects on soil and water quality result from increased recharge brought on by irrigation and clearance of land. These changes have caused enormous sections of land and water to become waterlogged and salinized in various parts of Australia, China, India, and the United States, making it impossible for adequate drainage to take place [46–49]. This process has a detrimental effect on agriculture, especially in low-lying or low-topographic relief locations where drainage is constrained, limiting the amount of arable land, reducing the growing season, and diminishing crop yields [50, 51].

On the other hand, the nature of recharge to underlying aquifers is considerably changed by urban contexts [52]. Groundwater resources are negatively impacted by the increased urbanization of areas in two fundamental ways: first, by preventing natural recharging of aquifers by covering the earth with concrete; second, by contaminating groundwater through drainage leaks and industrial waste and effluents [53]. Compared to clearing land for agriculture, the impact of urbanization on groundwater recharge is more complicated, and the impact on overall recharge rates varies depending on several site-specific factors related to the style and density of urban construction, as well as the type of infrastructure used to manage stormwater, sewage, and water supply [33, 54]. Urbanization always results in major changes to groundwater quality, recharge processes, and locations, although the total net change in recharge volume varies and is a subject of considerable ambiguity [55, 56]. It is well acknowledged that an increase in impermeable surfaces brought on by urbanization can locally diminish the rate of ground-water recharge and increase surface runoff, which then discharges to the urban drainage system [57, 58]. The excess run-off is often directed into storm-water management systems, including drains, pipelines, and retention basins along urban streams, even if pervious surfaces like roads and pavements may reduce groundwater recharge in their immediate proximity [59, 60].

Modifications to the current groundwater recharge process are frequently caused by unplanned urbanization and the increased strain that human activities are placing on hydro-geomorphologic systems [61, 62]. Anthropogenic activities are the primary cause of LULC alterations [63]. Numerous studies have been conducted utilizing remote sensing (RS) and geographic information systems (GIS) to evaluate LULC change and its effects on groundwater quality and quantity [12, 64–66]. Due to its extensive geographical and temporal coverage, remote sensing is a crucial technique for the investigation of LULC changes [65–68]. In addition, a variety of climatic factors also alter as urbanization replaces natural vegetation [69–71]. According to Kalnay and Cai [72], changes in land cover in the USA caused a rise in both the minimum and maximum temperatures. Additionally, groundwater condition (both quality and quantity) and its recharge are negatively impacted by urbanization [73, 74]. The hydrology of the region has been shown to have changed as a result of the conversion of natural, agricultural, and other low-population density sites into urban populations [75]. Evidence shows that when urbanization is excessive, more than half of the precipitation drains off and just a small portion is infiltrated deeply [76].

Based on the literature analysis mentioned above, it is clear that urbanization and the resulting changes in LULC have a negative impact on the local groundwater resources. However, the majority of the research to date has concentrated on analyzing bivariate correlations between urbanization and LULC changes [77, 78], urbanization and temperature changes [29, 79], urbanization and rainfall changes [80, 81], or urbanization and changes in groundwater level [82, 83]. Gebere et al. [84] investigated the effects of the dry Lake Haramaya watershed, located in the eastern region of Ethiopia. The simulated effects of future LULC were investigated using the land use change model CLUE-S (Conversion of Land Use and its Effects at Small regional extent). The WetSpass water balance model's simulated results showed that changes in land use and land cover had a significant impact on groundwater recharge in the watershed. In 2011, the yearly groundwater recharge varied from 0 to 90 mm. The range of recharge values reduced to 0–83 and 0–87 mm, respectively, according to a land use and land cover prediction made to the year 2028 under baseline and excellent management scenarios. The level of groundwater will also keep dropping due to increasing abstraction at the same time. Using an empirical method, Patra et al. [85] investigated the effects of urbanization on groundwater resources in the Howrah Municipal Corporation (HMC) in the Indian state of West Bengal. The outcome showed signs of urban sprawl or shrinking, which indicates an increase in a built-up area and leading to environmental degradation and groundwater contamination in the urban area. In the following sections, the impact of LULC on various aspects of groundwater resources is discussed in detail.

2.1 Impact on recharge

It is evident that the change in land cover has a significant impact on the change in groundwater recharge. Estimating groundwater recharge is crucial in managing water resources, especially in regions where groundwater is essential for the local water supply. According to Healy [86], groundwater recharge is defined as the vertical flow of water that reaches the water table and increases groundwater storage. Rates of recharge vary by orders of magnitude over space and time, depending on the interaction of climate, terrain, surface hydrology, vegetation, and land use [87, 88]. LULC influences the groundwater recharge variations over the spatial and temporal scales significantly [89]. Groundwater recharge is a crucial water balancing concept that is necessary to determine sustainable extraction rates and analyze aquifer sensitivity to pollution. It plays a significant role in regulating groundwater supply [33, 90]. Globally, the quantity, locations, and timing of groundwater recharge and discharge are increasingly altered due to rising population, agricultural growth, and urban land area. For groundwater development and sustainable groundwater resource management, groundwater recharge determines the groundwater withdrawal rates in a region [91].

Groundwater recharge, which occurs primarily through rainfall-recharge and surface water and groundwater interaction processes, replenishes groundwater aquifer systems. The change in LULC impacts groundwater recharge processes by modifying the earth's hydrological system functions. LULC consists of several subcategories as presented in **Figure 1**, each of which has particular effects on groundwater recharge. For instance, barren land in more populated areas inhibits groundwater penetration and lowers the pace at which groundwater systems recharge. On the other hand, based on other research, it was found that the decrease in evapotranspiration caused by surface sealing will boost groundwater recharge rates [57, 92]. For instance, in Austin, Texas, for example, [92] found that the groundwater recharge rate for the year 2000 was nearly twice as high as the pre-urban rate due to the contribution of urban recharge sources like water main leaks and excessive irrigation of, for example, gardens and agricultural areas. Similarly, in Perth, Australia, the practice of infiltrating roof and road runoff, along with decreased evaporative losses brought on by the growth of impermeable surfaces, leads to groundwater recharge rates 2–3 times greater than in pre-urban circumstances, according to research by Barron et al. [57].

Additionally, concrete structures caused by streets and buildings, flood control, forest management, and irrigation are examples of manmade activities that alter the infiltration and transport of water [93]. Groundwater supplies frequently deteriorate because of these alterations [34, 94]. Compared to the natural condition, urbanization alters the sources and flow routes of groundwater recharge [95]. Two urbanization-related processes have an impact on groundwater recharge on a quantitative level: (i) the growth of impervious surfaces, which reduces evapotranspiration and increases runoff [96]; and (ii) the construction of water supply and sewer networks, which boosts groundwater recharge rates because of leaks [97]. Compared to natural landscapes, urban environments' recharge declines since surface sealing limits infiltration and increases surface water runoff [98, 99]. According to Rose and Peters [100] study, urban wells' water levels noticeably dropped when compared to nonurban wells in the study region near Atlanta, US. In Dresden, Germany, Grischek et al. [98] discovered a 23% reduction in groundwater recharge as a result of surface sealing. Overall, all studies noted that because every city has a unique environment and frequently a distinct climate, therefore, it is challenging to anticipate the overall impact of urbanization on groundwater recharge.

Besides, urbanization and urban structure, natural vegetation has also various degrees of impact on groundwater recharge. For example, deep-rooted vegetation, like woods, has a lower rate of groundwater recharge than shallow-rooted vegetation, such as annual crops, according to Geist and Lambin [101]. Increased groundwater recharge rates are shown when natural deep-rooted native vegetation, such as trees and bushes, are replaced with shallow-rooted agricultural crops, based on field data [25, 47, 48] and modeling results [102]. But if the conversion of natural forests to cultivated crops lowers evapotranspiration losses, surplus water is available for boosting groundwater recharge and streamflow [24, 88]. In the past several decades, 80% of the woods in southwest Niger, which has a semiarid to an arid environment, have been turned into agriculture, increasing recharge from 2 to 25 + −7 mm year-1 [103]. For areas like these, where water is a limiting issue for sustainable development, it is crucial to look at the connections between LULC and groundwater change.

Similarly, the impact of deforestation on groundwater recharge has also been reported by many studies [104]. Deforestation increased the recharge and deep drainage by 1–2 times in Argentina [105], where the transition from grasslands to trees caused a 38 cm decrease in the water table [106]. Consequently, converting agricultural land back to "natural" vegetation may result in lower runoff and decreased in-stream sediment loads owing to reduced erosion, all essential components for sustainable water resource management [25, 35]. According to Brown et al., conversions to forests have decreased streamflow, changed the hydraulic characteristics of the soil, decreased soil moisture, and decreased recharge rates [104]. The considerably greater evapotranspiration rates of the planted woody plants are responsible for the decreases in groundwater recharge and soil moisture loss [107]. Other studies have also connected the decreases in soil moisture and recharge rates to vegetation-induced soil water repellency and increased rainfall interception of the plants that were planted [108–110].

For the purpose of assessing groundwater recharge, a number of techniques exist. They are roughly divided into physical, chemical, tracer, and numerical modeling techniques [111]. Studies have used a variety of methodologies to estimate groundwater recharge, including tracer methods, methods based on changes in the water table, lysimeter methods, and straightforward water balance procedures. In some of this research, recharge is incorporated into numerical groundwater models or dynamically linked to hydrological models to assess fluctuations under various climatic and land cover conditions [112, 113]. Lvovich [114] made the first effort at a world-scale study by producing a global recharge map using baseflow generated from river discharge hydrographs. The second large-scale groundwater recharge estimate was made by Döll [115], who used the WaterGAP Global Hydrological Predict [115, 116] to model global groundwater recharge at a spatial resolution of 0.50.

The impact of LULC on groundwater recharge has been monitored and studied using GIS and remote sensing data. Manap et al. [117] calculated the groundwater potential index in Malaysia using a number of geographical layers and remote-sensing photos. Based on the study of multi-temporal satellite and field survey data, Verma et al. [118] evaluated the effects of LULC on quaternary aquifer groundwater supplies in the Lucknow region of the Ganga plain, India. The findings showed that during the 10 years period, changes in LULC, hydro-geomorphic characteristics, and widespread groundwater research methods had led to significant changes in groundwater reservoirs.

Using remote sensing and GIS techniques, Waked et al. [83] investigated the effects of urbanization on groundwater recharge in the city of Hyderabad, India. According to the findings, the urban component of groundwater recharge was more than 10 times bigger than the natural component. Tam et al. [119] used a coupled hydrological simulation of rainfall-runoff and groundwater flow with WetSpa and MODFLOW to investigate the effects of urbanization on groundwater resources in Hanoi, Vietnam. According to the simulation's findings, seepage from rivers and lakes makes up 31% of the recharge of Hanoi City's groundwater system, while infiltration from rainfall provides 53.6%. The municipal water supply and sewage networks were responsible for the remaining 15.4% of the leakage. In the northwest of Bangladesh, Siddik et al. [120] investigated the impact of LULC changes on groundwater recharge. A semi-physically based water balance model was used to simulate spatially dispersed monthly groundwater recharge. The findings indicate that over the research period, the impervious built-up area rose by 80.3% while the vegetated land cover dropped by 16.4%. Because of this, groundwater recharge in 2016 was lower than it was in 2006. However, the reduction in recharge brought on by long-term temporal LULC changes is extremely negligible at the basin size (2.6 mm/year), even if urbanization has a bigger influence at the regional level (17.1 mm/year). Zomlot et al. [121, 122] used a change trajectory technique to examine the effects of land use change on groundwater recharge in Flanders, Belgium, and identified spatiotemporal LULC change trajectories.

The aforementioned studies have made significant headway in understanding the impact of various ULC types on groundwater resources however, there is a pertinent need to explore an in-depth analysis of the impact of grassland conversions to forest and grassland conversions to agricultural land. Similarly, the groundwater management plan and afforestation efforts, and future forest restoration plans around the globe need to be addressed in the current and future studies. Currently, significant efforts are being made to recover degraded forests in the United States [24, 104], China [95, 107], and India [123].

2.2 Impact on quality and quantity

Since LULC occurs globally and affects both water quantity and quality, its effects on groundwater are crucial [24]. Studies around the world have reported LULC's impact on groundwater recharge [124]. According to Valle et al. [125], groundwater quality changes are caused by people's direct or indirect associations with specific land uses. Numerous research has investigated the connection between LULC and groundwater contamination during the last few decades [126–128]. Barber et al. evaluated the effects of urbanization on groundwater quality related to LULC change using GIS-based techniques [1996]. Low NO3-N concentrations were found in forests and natural waterways and high NO3-N concentrations in croplands, according to research by Liu et al. [129] on the association between groundwater NO_3-N pollution and LULC types. The findings of Khan and Jhariya's [130] evaluation of the effect of LULC changes on groundwater quality using remote sensing, GIS, and field research revealed a 16.2% increase in the overall area of settlement from 1999 to 2016, which resulted in an increase in NO3 concentrations. He et al. [131] utilized the random forest (RF) method to forecast groundwater NO_3 concentrations in the Yinchuan region and concluded that the primary LULC categories influencing groundwater NO_3 concentrations were urban and agriculture.

GIS has been extensively used to study the effects of urban and land development on groundwater quality [132]. The geographical relationship between LULC changes and trends in groundwater quality has been explored by numerous studies [133, 134]. Most of these studies have been conducted to evaluate the effects of fast changes in LULC mapped using manual screen digitizing, which introduces bias and is prone to subjectivity [135]. Singh et al. [136] examined the effects of LULC in the lower Shiwalik hills in Rupnagar, Punjab, India, with a focus on groundwater quality and quantity. They discovered that changes in the LULC pattern led to an increase in groundwater quantity through both natural and artificial recharge. Due to the use of fertilizers intended to increase short-term soil fertility, the quality of groundwater has declined.

In the Pearl River Delta of China, urbanization coupled with the infiltration of domestic sewage was one of the main driving forces for groundwater quality in fissured aquifers in urbanized and peri-urban areas. Industrialization coupled with the infiltration of industrial wastewater was one of the main driving forces for groundwater quality in granular and fissured aquifers in peri-urban areas. Using a fuzzy synthetic assessment approach, it was determined that 83% of groundwater was drinkable with excellent quality. Compared to granular and fissured aquifers, groundwater in karst aquifers was drinkable and of higher quality. Groundwater quality in non-urbanized areas of the latter two aquifer types was much higher than that in peri-urban and urbanized areas [137].

According to Alqurashi and Kumar [138], recent fast urban growth has changed the original LULC patterns, modifying the groundwater circulation system and reducing groundwater quality owing to the leaching of contaminants from numerous sources (such as wastewater and fertilizers) [57, 139–141]. According to Townsend and Young [142], human causes, including urban and agricultural activities, were responsible for the groundwater nitrate (NO3-N) concentrations above 3 mg/L recorded in Kansas. On the other hand, urbanization, and industry greatly enriched sulfates (SO42), chlorides (Cl), and fluorides (F) in the shallow aquifers of Punjab, Pakistan, and Kharkiv, Ukraine [143, 144]. Elmahdy and Mohamed [145] discovered a connection between groundwater Cl, sodium (Na), and NO3 concentrations and animal waste, specifically from poultry.

Xu et al. [146] used statistical models and a curved streamline searchlight-shaped model to explore the geographical distribution patterns of groundwater hydrochemical parameters in the Guanzhong Basin and evaluate the correlations between the groundwater parameters and LULC (CS-SLM). The findings demonstrated that the north of the plain had greater groundwater parameter concentrations than the south. Most hydrochemical parameters, such as Na^{+}, Cl^{-}, SO_4^{2-}, F^{-}, and Cr^{6-}, were positively influenced by forests and water bodies, while negatively impacted by barren land and crops. Using Landsat pictures and hydrological data in a GIS, Elmahdy et al. [147] investigated the effects of LULC on groundwater level and quality in the northern United Arab Emirates. The findings demonstrated that the groundwater quality and level depletion across the research region, from the Oman Mountains to the coastal areas, is strongly correlated with the observed variations in LULC.

Harrington et al. and Xu et al. [23, 148] looked at the important connections between groundwater level decline and land use change. According to their findings, there is a clear connection between agricultural land-use change and groundwater depletion. In the Southern US, [24] looked at how LUCC affected groundwater recharge and soil quality. Their findings demonstrated the need for a quantitative understanding of the relationships between land change and groundwater recharge for sustainable land use. The substantial association between changes in land use and groundwater level was demonstrated by Chen et al. [149] using statistical analysis. Excessive aquifer withdrawal can have long-term effects on the land surface and the amount of groundwater.

The above studies depicted that, fertilizers, sewage disposal, and landfills are providing a possible source of groundwater pollution along with population growth and intensive agricultural practices, frequently affecting the groundwater quality. However, due to the wide variation in pollution concentrations throughout the landscape, it is challenging to measure and identify the relationship between human activities and groundwater quality and land use changes [13]. Also, there is a lack of assessment of the drivers of groundwater pollution in the transboundary groundwater system and land use conflicts on the quality of groundwater [125].

2.3 Impact on groundwater storage

Groundwater storage is another factor that is affected by the change in LULC. Various studies have assessed the impact of LULC on groundwater storage [11, 150, 151]. Groundwater recharge, groundwater outflow, and groundwater level all show a positive spatial association with LULC variations. According to Haddeland et al. [151], the human population directly affects the terrestrial water cycle, thus impacting the groundwater resource.

Land factors and their non-hydrological equivalents, such as temperature and LULC, have an influence on groundwater storage. For instance, in the Ganga basin, rising demand from the agricultural and industrial sectors as well as population growth have had a significant impact on groundwater storage, geochemical properties, and the type and extent of water exchange with the river [152, 153]. The negative effects may be seen as a long-term decline in groundwater levels, desaturation of aquifer zones, higher energy need to raise water from deeper levels, and quality degradation brought on by saline water intrusion in coastal areas in various sections of the nation [154].

The effects of LULC variations on evapotranspiration and groundwater storage were examined by Dias et al. [150]. The increased impermeable surface has been a significant contributor to decreased infiltration, which leads to decreased groundwater storage [155]. Groundwater is a vital component that significantly affects the well-being of people, animals, and aquatic environments. However, anthropogenic activities, including intensive agricultural crop production, urbanization, mining, and industrial developments, pose significant pollution threats to the sustainability of groundwater resources [156]. The main factor for groundwater pollution is human activity. Because of the excessive use of pesticides, herbicides, and large-scale applications of nitrogenous fertilizer, more agricultural operations will have a greater probability of contaminating the groundwater [157].

A thorough understanding of groundwater inputs and outputs is needed to inform water management decisions for future planning and policy. Due to population increase and the effects it has on built-up expansion, impermeable surfaces impede groundwater recharge and other hydro-climatic factors [158, 159]. However, because of the impact of social, environmental, and economic issues, it is challenging to determine land-use trends [33]. Furthermore, combining knowledge and modeling capabilities across biophysical responses, environmental issues, policies, economies, data, and computer capabilities is necessary to fully understand the long-term effects of natural and human causes in river basin interactions [160].

3. Complexity of groundwater management

Groundwater supports drinking water for the population and irrigation for agriculture. However, the excessive use of groundwater resources has become a major issue on a worldwide scale and needs an immediate response [161, 162]. Despite being a vitally significant global water resource, groundwater receives less attention from management systems than readily accessible surface water resources, as stated by Famiglietti [163]. This is especially true in nations with poor or nonexistent water administration and insufficient aquifer monitoring. As a result, there is a paucity of information about (i) how groundwater storage responds to different drivers such as point sources of pollution (ii) how storage variations relate to aquifer heterogeneity and transboundary aquifer, and (iii) how future changes in groundwater levels may be expected and mitigated based on these different drivers. Such knowledge is essential for management, particularly if the heterogeneity of the aquifer system causes noticeably varied reactions to future stresses in various sections of a region.

Assessing LULC change and its effects on groundwater quality have been done extremely effectively using remote sensing and GIS [164–166]. The capacity of remote sensing data to provide information on geographical and temporal domains, which is crucial for effective analysis and prediction, is one of the biggest benefits of employing it for hydrogeological research and monitoring [61, 166]. Further research needs to be done, to collect greater precision data. Therefore, there is a need to integrate remote sensing photos, field surveys, and visual interpretations to collect more precise data and further examine the danger of groundwater contamination influenced by LULC in future studies.

The absence of surface water monitoring is lacking which typically helps to constrain catchment water balances and improve recharge estimates. It is crucial to gather a range of independent field data to support and improve conceptual models and measure groundwater recharge [167]. Without such information, mapping, and analysis, it would be challenging to identify the spatial dependencies and major variables influencing recharge, which would leave room for uncertainty regarding how future water budgets and water quality may change as land-use change occurs.

Evaluating changes due to LULC, and the consequences for processes like aquifer depletion, land subsidence, and land and water salinization, are important scientific challenges resulting from the complexity of recharging processes and difficulty in accurately estimating recharge even in relatively undisturbed environments [168, 169]. Further, things are complicated by time lags that are typical of groundwater systems' reactions to hydrological change at the surface [170–172], as well as feedback between changes in recharge and other elements of the eco-hydrological system [173]. To successfully understand, predict, and manage groundwater systems, analysis, and assessment of the effects of human-induced activities on recharge processes and rates are necessary [174, 175] because groundwater recharge negatively and positively impacts people's quality of life through a complex social-ecological hydrologic system [45]. Hence, it is essential for sustainable land and water management to have a thorough understanding of the connections between vegetation and groundwater recharge as well as the water balance consequences of agricultural land use. Moreover, due to increased groundwater recharge sources and widely dispersed new abstraction points in the urbanized area, the water balancing of an urban aquifer is complicated. Understanding the effects of changing land uses on groundwater recharge is crucial, particularly in areas where urbanization is taking place rapidly with limited surface water [176].

Due to the challenges of assessing groundwater recharge, previous studies on groundwater recharge have concentrated on limited points in space [177]. In order to quantify the spatiotemporal variability of groundwater recharge, numerical modeling is an alternative [178, 179]. It is important to note that there were various levels of variance between studies which can be attributed to many factors. First, the major cause of the variance may be a mismatch in the spatial size. Although the simulation is point-based, the vegetation data were typical of a large pixel scale, therefore the simulated groundwater recharge should be evaluated in the context of a broad scale [107] Second, inconsistent temporal scales between studies may contribute to the underestimating of groundwater recharge [180]. Third, management techniques, such as irrigation and soil diversity, would also have an impact on the outcomes. Irrigated agriculture typically receives more water input than rainfed crops and produces greater groundwater recharge [181]. On the other hand, the influence of irrigation on groundwater recharge is less pronounced at the regional level if the irrigation water's primary source is the same region's rainfall, which is presumably the case in most cases.

Even though the loss of groundwater supplies is extremely localized, coarser satellite data like Gravity Recovery and Climate Experiment (GRACE) frequently mask it [182]. Because of the limitations of groundwater data and their coarse spatiotemporal resolution, it can be difficult to understand the spatiotemporal variability of groundwater. Studies of groundwater variability now rely on two primary data sources: (i) local *in situ* data from borehole data, such as those from [183]; and (ii) satellite-based GRACE data paired with hydrological or reanalysis models [33, 184, 185]. Likewise, the resulting groundwater outputs from the associated hydrological models are lowered due to the uncertainty of the meteorological forcing inputs [186]. For instance, mean annual precipitation in the southwest of the United States accounts for 80% of the variance in ground recharge [102, 103]. The impact of climatic conditions on groundwater recharge, however, varies greatly depending on the site [187]. The impact of meteorological elements can also change depending on the kind of aquifer, irrigation intensity, and seasonal variability of precipitation [188, 189]. Extreme precipitation was discovered to have a substantial influence in influencing groundwater recharge across the Northern High Plains in the United States, contrary to the conventional hypothesis that claims groundwater recharge was controlled by low-intensity precipitation over extended periods of time [190, 191]. It is unknown how the groundwater recharge dynamics were impacted by these temporal variations in the rainfall pattern [192]. Additionally, the seasonality of precipitation was said to have a big impact on groundwater recharge [193].

Groundwater, the biggest distributed reservoir of fresh water on Earth, is crucial for maintaining ecosystems and allowing for human adaptability to climate change [194]. In order to maintain a robust and sustainable economy in the future, an accurate evaluation of groundwater resources is essential. But groundwater supply and quality are influenced by a variety of social, economic, and environmental factors, and these systems are frequently unpredictable. Therefore, they make managing groundwater more difficult. Groundwater is used by populations in every part of the world to varying degrees. Groundwater is a complicated system that, once damaged, is particularly challenging to restore [195]. In addition, the effects of future land-use changes on the groundwater system have not been well studied. Groundwater extraction for irrigation will rise by 39% by 2050 [196], despite the fact that the world's population, Gross Domestic Product (GDP), and water demand would be unequal, posing difficulties to around 30% of the world's major groundwater systems [197].

Thus, there are several factors that influence LULC change and its impact on groundwater. These factors include difficult-to-trace socioeconomic and biophysical variables, which in turn lead to a complex and changing system [198, 199]. The applicability of any particular driving factor in a particular circumstance relies on the social and geographic context. A clear division of the multiple causes is frequently challenging, given the interconnected impacts of climate and LULC change [200]. Regarding the economic drivers, it is hypothesized that urbanization and the spread of agricultural and grasslands provide advantageous economic and institutional conditions that lead to land use change. In political ecology, it is believed that the maintenance of globalization, the market forces of capital, and multinational corporations are frequently what drive changes in land usage. Possible drivers of land use change include the creation of unsustainable irrigation systems where groundwater exploration exceeds recharge rates [201], the appropriation of land and water [202], and the distribution of land to elites that marginalizes the disadvantaged communities [198].

Moreover, underlying causes of land cover change do not take place in isolation; they involve intricate interactions across these several scales [203]. Social, political, economic, demographic, technical, cultural, institutional, and biophysical variables are among the root causes of changes in land cover [204, 205]. Land use change may happen gradually or even more quickly as a result of certain occurrences like natural disasters or shifts in political power [206]. It is important that these data may be studied in the context of how local land use decisions related to both cultural and politico-economic forces, and how land cover patterns are impacted by such processes. Physical landscapes and social systems are as much the result of uneven power relations, histories of colonialism, and racial and gender inequality as they are of hydrology, ecology, and climate change, so we cannot rely on explanations based just on physical or critical human geography [207].

In order to resolve these problems, a coupled sociocultural and natural systems approach as proposed in **Figure 2**, as an example, is required to enable efficient interaction between social scientists, biophysical scientists, and management experts and to better comprehend how individuals interact with their surroundings to impact groundwater. Untangling the intricacies of linked human and natural systems, such as reciprocal effects, the impact of many scales of biological and social organization,

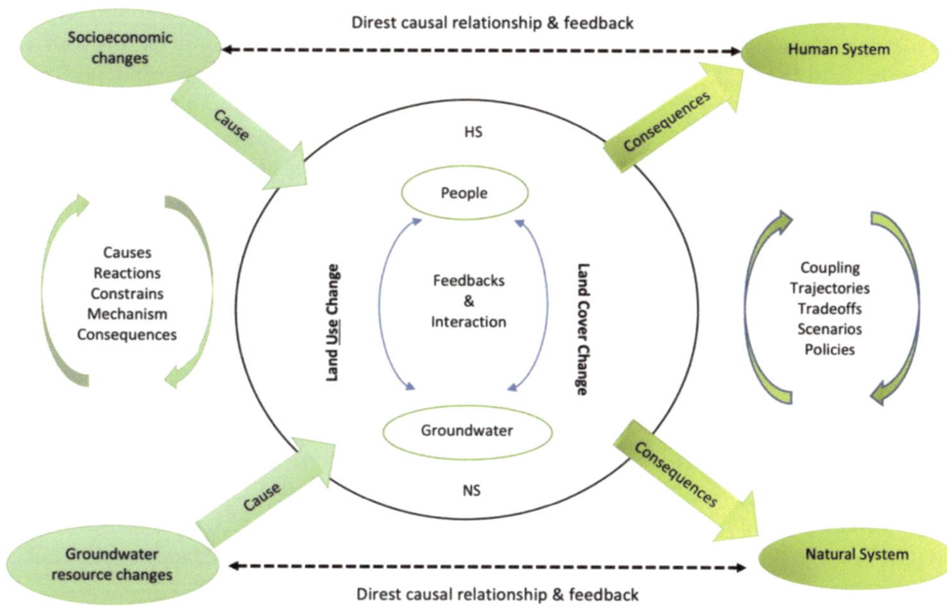

Figure 2.
Coupled human-natural conceptual framework to study groundwater (adapted from [80, 81]).

and emergent features can result in innovative scientific findings that are crucial for the creation of successful regulations for ecological and socioeconomic sustainability [208]. When human and natural systems are investigated together, new and complex patterns and processes emerge that are not visible when the two disciplines are examined individually [209]. Opportunities to properly combine different disciplines are becoming available in order to address basic coupled sociocultural and natural systems issues and respond to society's tremendous problems [209]. In other words, groundwater resources management challenges should be identified and addressed through multidisciplinary studies and initiatives. The relatively new science of coupled human-natural systems offers a promising framework to tackle the complex problems of groundwater resource management by recognizing the integrated and coupled nature of human and ecological systems [210].

This conceptual framework encourages synthesizing research approaches to foster innovative studies that advance our understanding of the complex socio-biophysical phenomena and develop socially and environmentally resilient policy outcomes for sustainable groundwater management. It will further help improve our understanding of the cause, exposures, and consequences of groundwater issues and devise appropriate strategies to combat future groundwater depletion and improvement of lives of the people. The proposed framework offers a possible course of action as an example. It can be modified to best suit the needs of a specific study and objectives.

4. Conclusion

The sustainability of groundwater resources is crucial for the areas with the highest population growth, particularly the arid and semiarid areas that depend on the resource for domestic, industrial, and agricultural requirements. Half of the world's population relies on groundwater as their primary drinking water source. It also maintains ecosystems by providing them with access to water, nutrients, and a reasonably stable temperature. Therefore, groundwater contribution is crucial for most regions around the globe [211].

LULC change is becoming a major ecological hazard, particularly the conversion of natural vegetation into croplands or the degradation of land into barren. Intense human exploitation of land resources throughout history has led to considerable changes in land use and cover. The phenomenon of LULC has significantly intensified in many locations since the age of industrialization and high population expansion. As a result, the demand for groundwater resources is expanding, and the amount of water available per person is decreasing every day as a result of our population's rapid growth and rising standards of living.

The growth in groundwater abstraction to meet the demand for water supply from an increasingly urban population, on the other hand, is one of the reasons leading to groundwater depletion around the world. To fulfill the demands of domestic, commercial, industrial, and public users, water must be supplied. LULC change influences the volume, forms, and patterns of groundwater recharge [89]. Such a change might impact the environment and the socioeconomic condition in a number of different ways, both directly and indirectly [33].

For the sustainable use of groundwater resources, as well as the adaptation and mitigation of climate change, LULC must be managed. To achieve this, an integrated approach to groundwater recharge is necessary to ensure a link between recharge and abstraction and to comprehend the impact of LULC and climate change on the spatiotemporal distribution of recharge [121, 122, 212, 213]. Understanding the dynamics and causes of groundwater recharge is essential for forecasting and managing groundwater systems as well as the ongoing expansion of the water supply [33, 214].

Author details

Shobha Kumari Yadav
West Virginia University, Morgantown, USA

*Address all correspondence to: shobhas.yadav@gmail.com; sky0003@mix.wvu.edu

References

[1] Lall U, Josset L, Russo T. A snapshot of the world's groundwater challenges. Annual Review of Environment and Resources. 2010;**45**:171-194. DOI: 10.1146/annurev-environ-102017-0258

[2] Shahid S, Wang X-J, Rahman MM, Hasan R, Harun SB, Shamsudin S. Spatial assessment of groundwater over-exploitation in northwestern districts of Bangladesh. Journal of the Geological Society of India. 2015;**85**:463-470

[3] Siebert S, Burke J, Faures JM, Frenken K, Hoogeveen J, Döll P, et al. Groundwater use for irrigation—A global inventory. Hydrology and Earth System Sciences. 2010;**14**:1863-1880

[4] Sharma KD. Groundwater management for food security. Current Science. 2009;**96**:1444-1447

[5] Argent R. Inland water: Groundwater resources, Australia state of the environment 2016. In: Tech. Rep. Australian Government Department of the Environment and Energy. Canberra; 2016. DOI: 10.4226/94/58b656cfc28d1

[6] Jasechko S, Perrone D, Befus KM, Bayani Cardenas M, Ferguson G, Gleeson T, et al. Global aquifers dominated by fossil groundwaters but wells vulnerable to modern contamination. Nature Geoscience. 2017;**10**:425-429

[7] Scanlon BR, Faunt CC, Longuevergne L, Reedy RC, Alley WM, McGuire VL, et al. Groundwater depletion and sustainability of irrigation in the US High Plains and Central Valley. Proceedings of the National Academy of Sciences. 2012;**109**:9320-9325

[8] Wada Y, van Beek LPH, Bierkens MFP. Nonsustainable groundwater sustaining irrigation: A global assessment. Water Resources Research. 2012;**48**:335-344

[9] Ojeda Olivares EA, Sandoval Torres S, Belmonte Jiménez SI, Campos Enríquez JO, Zignol F, Reygadas Y, et al. Climate change, land use/land cover change, and population growth as drivers of groundwater depletion in the central valleys, Oaxaca, Mexico. Remote Sensing. 2019;**11**(11):1290. DOI: 10.3390/rs11111290

[10] Greenbaum D. Structural influences on the occurrence of groundwater in SE Zimbabwe. In: Wright EP, Burges WG, editors. Hydrology of Crystalline Basement Aquifers in Africa. Vol. 66. Geol. Soc. Spec. Publ., Bath; 1992. pp. 77-85

[11] Mukherjee P, Singh KC, Mukherjee S. Delineation of groundwater potential zones in arid region of India-a remote sensing and GIS approach. Water Resource Management. 2012;**26**:2643-2672

[12] Singh KC, Shashtri S, Mukherjee S, Kumari R, Avatar R, Singh A, et al. Application of GWQI to Assess Effect of Land Use Change on Groundwater Quality in Lower Shiwaliks of Punjab: Remote Sensing and GIS Based Approach. Vol. 252011. pp. 1881-1898

[13] Zhang X, Xu Z, Sun X, Dong W, Ballantine D. Nitrate in shallow groundwater in typical agricultural and forest ecosystems in China, 2004-2010. Journal of Environmental Sciences. 2013;**25**(5):1007-1014. DOI: 10.1016/S1001-0742(12)60139-9

[14] Wada Y, Wisser D, Bierkens MF. Global modeling of withdrawal, allocation and

consumptive use of surface water and groundwater resources. Earth System Dynamics. 2014;**5**(1):15-40

[15] Dasgupta B, Sanyal P. Linking Land Use Land Cover change to global groundwater storage. Science of the Total Environment. 2022;**853**:158618. DOI: 10.1016/j.scitotenv.2022.158618

[16] Shrestha S, Semkuyu DJ, Pandey VP. Assessment of groundwater vulnerability and risk to pollution in Kathmandu Valley, Nepal. Science of the. Total Environment. 2016;**556**:23-35

[17] Mahmon NA, Ya'acob N, Yusof AL. Differences of image classification techniques for land use and land cover classification. In: IEEE 11th International Colloquium on Signal Processing & its Applications (CSPA). IEEE; 2015. pp. 90-94

[18] Sun J, Yu X, Wang H, Jia G, Zhao Y, Tu Z, et al. Effects of forest structure on hydrological processes in China. Journal of Hydrology. 2018;**561**:187-199. DOI: 10.1016/j.jhydrol.2018.04.003

[19] Wang G, Li J, Sun W, et al. Non-point source pollution risks in a drinking water protection zone based on remote sensing data embedded within a nutrient budget model. Water Research. 2019;**157**:238-246. DOI: 10.1016/j.watres.2019.03.070

[20] Richards WF. Land transformation. In: Turner BL, Clark WC, Kates RW, et al., editors. The Earth as Transformed by Human Action. Cambridge: Cambridge University Press; 1990. pp. 561-575

[21] Agam N, Evett SR, Tolk JA, Kustas WP, Colaizzi PD, Alfieri JG, et al. Evaporative loss from irrigated interrows in a highly advective semi-arid agricultural area. Advances in Water Resources. 2012;**50**:20-30

[22] Allen RG, Pereira LS, Raes D, Smith M. Crop Evapotranspiration – Guidelines for Computing Crop Water Requirements FAO Irrigation and Drainage Paper 56. Rome, Italy: FAO; 1998

[23] Harrington L, Harrington J Jr, Kettle N. Groundwater depletion and agricultural land use change in the high plains: A case study from Wichita County, Kansas. The Professional Geographer. 2007;**59**:221-235

[24] Scanlon BR, Reedy RC, Stonestrom DA, Prudic DE, Dennehy KF. Impact of land use and land cover change on groundwater recharge and quality in the southwestern US. Global Change Biology. 2005;**11**:1577-1593. DOI: 10.1111/j.1365-2486.2005.01026.x

[25] Scanlon BR, Jolly I, Sophocleous M, Zhang L. Global impacts of conversions from natural to agricultural ecosystems on water resources: Quantity versus quality. Water Resources Research. 2007;**43**(3)

[26] Turner SW, Hejazi M, Calvin K, Kyle P, Kim S. A pathway of global food supply adaptation in a world with increasingly constrained groundwater. Science of the Total Environment. 2019;**673**:165-176

[27] Albhaisi M, Brendonck L, Batelaan O. Predicted impacts of land use change on groundwater recharge of the upper Berg catchment. South Africa Water SA. 2013;**39**:211-220

[28] Batelaan O, De Smedt F, Triest L. Regional groundwater discharge: Phreatophyte mapping, groundwater modelling and impact analysis of land-use change. Journal of Hydrology. 2003;**275**:86-108. DOI: 10.1016/S0022-1694(03)00018-0

[29] Wang J, Yan Z, Li Z, Liu W, Wang Y. Impact of urbanization on changes in temperature extremes in Beijing during

1978-2008. Chinese Science Bulletin. 2013;**58**(36):4679-4686

[30] Wada Y, van Beek LPH, van Kempen CM, Reckman JWTM, Vasak S, Bierkens MFP. Global depletion of groundwater resources. Geophysical Research Letters. 2010;**37**:L20402

[31] Meyer B, Turner I, editors. Changes in Land Use and Land Cover, a Global Perspective. Cambridge: Cambridge University Press; 1994 537 P

[32] Green TR, Taniguchi M, Kooi H, Gurdak JJ, Allen DM, Hiscock KM, et al. Beneath the surface of global change: Impacts of climate change on groundwater. Journal of Hydrology. 2011;**405**:532-560

[33] Han D, Currell M, Cao G, Hall B. Alterations to groundwater recharge due to anthropogenic landscape change. Journal of Hydrology. 2017;**554**:545-557. DOI: 10.1016/j.jhydrol.2017.09.018

[34] Minning M, Moeck C, Radny D, Schirmer M. Impact of urbanization on groundwater recharge rates in Dübendorf, Switzerland. Journal of Hydrology. 2018;**563**:1135-1146

[35] Owuor SO, Butterbach-Bahl K, Guzha AC, Rufino MC, Pelster DE, Díaz-Pinés E, et al. Groundwater recharge rates and surface runoff response to land use and land cover changes in semi-arid environments. Ecological Processes. 2016;**5**(1):1-21

[36] Bundschuh J, Farias B, Martin R, Storniolo A, Bhattacharya P, Cortes J, et al. Groundwater arsenic in the Chaco-Pampean plain, Argentina: Case study from robles county, Santiago del Estero province. Applied Geochemistry. 2004;**19**(2):231-243

[37] Srivastava PK, Singh SK, Gupta M, Thakur JK, Mukherjee S. Modeling impact of land use change trajectories on groundwater quality using remote sensing and GIS. Environmental Engineering and Management Journal. 2013;**12**(12)

[38] Vörösmarty C, Lettenmaier D, Leveque C, Meybeck M, Pahl-Wostl C, Alcamo J, et al. Humans transforming the global water system. EOS. Transactions of the American Geophysical Union. 2004;**85**(48):509-514

[39] Chhabra A, Geist H, Houghton RA, Haberl H, Braimoh AK, Vlek PLG, et al. Multiple impacts of land-use/cover change. In: Land-Use and Land-Cover Change: Local Processes and Global Impacts. Berlin: Springer; 2006

[40] Stonestrom DA, Scanlon BR, Zhang L. Introduction to special section on impacts of land use change on water resources. Water Resources Research. 2009;**45**:1-3

[41] Dams J, Woldeamlak ST, Batelaan O. Predicting land-use change and its impact on the groundwater system of the Kleine Nete catchment, Belgium. Hydrology and Earth System Sciences. 2008;**12**:1369-1385

[42] Mkaya DM, Mutua BM, Kundu PM. Evaluation of the impact of land use change on catchment hydrology: The case of Wundanyi River catchment in Taita hills, Kenya. Research Journal of Agricultural and Environmental Management. 2013;**2**(5):92-98

[43] Eckhardt K, Ulbrich U. Potential impacts of climate change on groundwater recharge and streamflow in a central European low mountain range. Journal of Hydrology. 2003;**284**(1):244-252

[44] Allison GB, Hughes MW. The use of natural tracers as indicators of soil-water movement in a temperate semi-arid region. Journal of Hydrology. 1983;**60**:157-173

[45] Booth EG, Zipper SC, Loheide SP II, Kucharik CJ. Is groundwater recharge always serving us well? Water supply provisioning, crop production, and flood attenuation in conflict in Wisconsin, USA. Ecosystem Services. 2016;**21**:153-165

[46] Rengasamy P. World salinization with emphasis on Australia. Journal of Experimental Botany. 2006;**57**:1017-1023

[47] Allison GB, Cook PG, Barnett SR, Walker GR, Jolly ID, Hughes MW. Land clearance and river salinisation in the western Murray Basin, Australia. Journal of Hydrology. 1990;**119**:1-20

[48] Allison GB, Cook PG, Barnett SR, Walker GR, Jolly ID, Hughes MW. Land clearance and river salinization in the western Murray Basin Australia. Journal of Hydrology. 1990;**119**:1-20

[49] Ghassemi F, Jakeman AJ, Nix HA. Salinisation of Land and Water Resources: Human Causes, Extent, Management and Case Studies. Wallingford: University of New South Wales Press, Sydney and CAB International; 1995

[50] Hughes AG, Vounaki T, Peach DW, Ireson AM, Jackson CR, Butler AP, et al. Flood risk from groundwater: Examples from a chalk catchment in southern England. Journal of Flood Risk Management. 2011;**4**:143-155

[51] Zipper SC, Soylu ME, Booth EG, Loheide SP II. Untangling the effects of shallow groundwater and soil texture as drivers of subfield-scale yield variability. Water Resources Research. 2015;**51**:6338-6358

[52] Zipper SC, Soylu ME, Kucharik CJ, Loheide SP II. Quantifying indirect groundwater-mediated effects of urbanization on agroecosystem productivity using MODFLOW-agro IBIS (MAGI), a complete critical zone model. Ecological Modelling. 2017;**359**:201-219

[53] Baier K, Schmitz KS, Azzam R, Strohschon R. Management tools for sustainable ground water protection in mega urban areas-small scale land use and ground water vulnerability analyses in Guangzhou, China. International. Journal of Environmental Research. 2014;**8**:249-262

[54] Locatelli L, Mark O, Mikkelsen PS, Arnbjerg-Nielsen K, Deletic A, Roldin M, et al. Hydrologic impact of urbanization with extensive storm water infiltration. Journal of Hydrology. 2017;**544**:524-537

[55] Mapani BS. Groundwater and urbanisation, risks and mitigation: The case for the city of Windhoek, Namibia. Physics and Chemistry of the Earth. 2005;**30**:706-711

[56] Tubau I, Vázquez-Suñé E, Carrera J, Valhondo C, Criollo R. Quantification of groundwater recharge in urban environments. Science of the Total Environment. 2017;**592**:391-402

[57] Barron OV, Barr AD, Donn MJ. Evolution of nutrient export under urban development in areas affected by shallow water table. Science of the Total Environment. 2013;**443**:491-504

[58] Bhaduri B, Harbor J, Engel B, Grove M. Assessing watershed-scale, long- term hydrologic impacts of land-use change using a GIS-NPS model. Environmental Management. 2000;**26**(6):643-658

[59] Chen J, Theller L, Gitau MW, Engel BA, Harbor JM. Urbanization impacts on surface runoff of the contiguous United States. Journal of Environmental Management. 2017;**187**:470-481

[60] Epting J, Huggenberger P, Rauber M. Integrated methods and scenario development for urban groundwater management and protection

during tunnel road construction: A case study of urban hydrogeology in the city of Basel, Switzerland. Hydrogeology Journal. 2008;**16**:575-591

[61] Li S, Ma Y. Urbanization, economic development and environmental change. Sustainability. 2014;**6**:5143-5161

[62] Peng X, Chen X, Cheng Y. Urbanization and its consequences. Demography. 2009;**2**

[63] Yin J, Yin Z, Zhong H, Xu S, Hu X, Wang J, et al. Monitoring urban expansion and land use/land cover changes of Shanghai metropolitan area during the transitional economy (1979-2009) in China. Environmental Monitoring and Assessment. 2011;**177**:609-621

[64] Kim G. Integrated consideration of quality and quantity to determine regional groundwater monitoring site in South Korea. Water Resource Management. 2009;**24**(14):4009-4032

[65] Meera K, Slade L, Levine H. Solvent retention capacity (SRC) testing of wheat flour: Principles and value in predicting flour functionality in different wheat-based food processes and in wheat breeding—A review. Journal of Cereal Chemistry. 2011;**88**(6):537-552

[66] Singh NP, Mukherjee TK, Shrivastava BBP. Monitoring the impact of coal mining and thermal power industry on land use pattern in and around Singrauli coalfiled using remote sensing data and GIS. Journal of the Indian Society of Remote Sensing. 1997;**25**:61-72

[67] Anderson JR, Hardy EE, Roach JT, Witmer RE. A land use and land cover classification system for use with remote sensor data. USGS Prof. Paper. 1976;**964**:1-26

[68] Green K, Kempka D, Lackley L. Using remote sensing to detect and monitor land cover and land use changes. Photogrammetric Engineering and Remote Sensing. 1994;**60**(3):331-337

[69] Cui L, Shi J. Urbanization and its environmental effects in Shanghai, China. Urban Climate. 2012;**2**:1-15

[70] Kometa SS, Akoh NR. The hydro-geomorphological implications of urbanisation in Bamenda, Cameroon. Journal of Sustainable Development. 2012;**5**(6):64

[71] Zhao S, Da L, Tang Z, Fang H, Song K, Fang J. Ecological consequences of rapid urban expansion: Shanghai, China. Frontiers in Ecology and the Environment. 2006;**4**(7):341-346

[72] Kalnay E, Cai M. Impact of urbanization and land-use change on climate. Nature. 2003;**423**(6939):528-531

[73] Graniel CE, Morris LB, Carrillo-Rivera JJ. Effects of urbanization on groundwater resources of Merida, Yucatan, Mexico. Environmental Geology. 1999;**37**(4):303-312

[74] Karamouz M, Ahmadi A, Akhbari M. Groundwater Hydrology: Engineering, Planning, and Management. CRC Press; 2011

[75] Blanco H, McCarney P, Parnell S, Schmidt M, Seto KC. The role of urban land in climate change. In: Climate Change and Cities: First Assessment Report of the Urban Climate Change Research. Cambridge: Cambridge University Press; 2011. p. 240

[76] Arnold CL Jr, Gibbons CJ. Impervious surface coverage: The emergence of a key environmental indicator. Journal of the American Planning Association. 1996;**62**(2):243-258

[77] Alqurashi AF, Kumar L. An assessment of the impact of urbanization

and land use changes in the fast-growing cities of Saudi Arabia. Geocarto International. 2017;**34**:1-20

[78] Sajjad H, Iqbal M. Impact of urbanization on land use/land cover of Dudhganga watershed of Kashmir Valley, India. International Journal of Urban Sciences. 2012;**16**(3):321-339

[79] Chapman S, Watson JE, Salazar A, Thatcher M, McAlpine CA. The impact of urbanization and climate change on urban temperatures: A systematic review. Landscape Ecology. 2017;**32**(10):1921-1935

[80] Chen J, John R, Zhang Y, Shao C, Brown DG, Batkhishig O, et al. Divergences of two coupled human and natural systems on the Mongolian plateau. Bioscience. 2015;**65**:559-570. DOI: 10.1093/biosci/biv050

[81] Chen S, Li WB, Du YD, Mao CY, Zhang L. Urbanization effect on precipitation over the Pearl River Delta based on CMORPH data. Advances in Climate Change Research. 2015;**6**(1):16-22

[82] Khazaei E, Mackay R, Warner JW. The effects of urbanization on groundwater quantity and quality in the Zahedan aquifer, Southeast Iran. Water International. 2004;**29**(2):178-188

[83] Wakode HB, Baier K, Jha R, Azzam R. Impact of urbanization on groundwater recharge and urban water balance for the city of Hyderabad, India. International Soil and Water Conservation Research. 2018;**6**:51-62

[84] Gebere SB, Alamirew T, Merkel BJ, Melesse AM. Land use and land cover change impact on groundwater recharge: The case of Lake Haramaya watershed, Ethiopia. In: Melesse A, Abtew W, editors. Landscape Dynamics, Soils and Hydrological Processes in Varied Climates. Cham: Springer Geography; 2016. DOI: 10.1007/978-3-319-18787-7_6

[85] Patra S, Sahoo S, Mishra P, Mahapatra SC. Impacts of urbanization on land use/cover changes and its probable implications on local climate and groundwater level. Journal of Urban Management. 2018;**7**:70-84. DOI: 10.1016/j.jum.2018.04.006

[86] Healy RW. Estimating Groundwater Recharge. Cambridge University Press; 2010

[87] McMahon PB, Plummer LN, Böhlke JK, Shapiro SD, Hinkle SR. A comparison of recharge rates in aquifers of the United States based on groundwater-age data. Hydrogeology Journal. 2011;**19**:779-800

[88] Scanlon BR, Keese KE, Flint AL, Flint LE, Gaye CB, Edmunds WM, et al. Global synthesis of groundwater recharge in semiarid and arid regions. Hydrological Processes. 2006;**20**(15):3335-3370

[89] Ansari TA, Katpatal YB, Vasudeo AD. Spatial evaluation of impacts of increase in impervious surface area on SCS-CN and runoff in Nagpur urban watersheds India. Arabian Journal of Geosciences. 2016;**9**:702. DOI: 10.1007/s12517-016-2702-5

[90] Alley WM, Healy RW, LaBaugh JW, Reilly TE. Flow and storage in groundwater systems. Science. 2002;**296**(5575):1985-1990. DOI: 10.1126/science.1067123

[91] Mohan C, Western AW, Wei Y, Saft M. Predicting groundwater recharge for varying land cover and climate conditions-a global meta-study. Hydrology and Earth System Sciences. 2018;**22**:2689-2703. DOI: 10.5194/hess-22-2689-2018

[92] Garcia-Fresca B. Urban-enhanced groundwater recharge: Review and case

study of Austin, Texas, USA. In: Urban Groundwater, Meeting the Challenge: IAH Selected Papers on Hydrogeology 8. CRC Press; 2007

[93] Baillieux A, Moeck C, Perrochet P, Hunkeler D. Assessing groundwater quality trends in pumping wells using spatially varying transfer functions. Hydrogeology Journal. 2015;**23**:1449. DOI: 10.1007/s10040-015-1279-5

[94] Schirmer M, Leschik S, Musolff A. Current research in urban hydrogeology – A review. Advances in Water Resources. 2013;**51**:280-291. DOI: 10.1016/j.advwatres.2012.06.015

[95] Yang Y, Lerner DN, Barrett MH, Tellam JH. Quantification of groundwater recharge in the city of Nottingham, UK. Environmental Geology. 1999;**38**(3):183-198. DOI: 10.1007/s002540050414

[96] Paulachok GN. Geohydrology and Ground-Water Resources of Philadelphia, Pennsylvania. US Geological Survey Water-Supply Paper; 1991. p. 2346

[97] del Campo MAM, Esteller MV, Exposito JL, Hirata R. Impacts of urbanization on groundwater hydrodynamics and hydrochemistry of the Toluca Valley aquifer (Mexico). Environmental Monitoring and Assessment. 2014;**186**(5):2979-2999. DOI: 10.1007/s10661-013-3595-3

[98] Grischek T, Nestler W, Piechniczek D, Fischer T. Urban groundwater in Dresden, Germany. Hydrogeology Journal. 1996;**4**(1):48-63. DOI: 10.1007/s100400050088

[99] Hardison EC, O'Driscoll MA, DeLoatch JP, Howard RJ, Brinson MM. Urban land use, channel incision, and water table decline along coastal plain streams, North Carolina. JAWRA Journal of the American Water Resources Association. 2009;**45**(4):1032-1046. DOI: 10.1111/j.1752-1688.2009.00345.x

[100] Rose S, Peters NE. Effects of urbanization on streamflow in the Atlanta area (Georgia, USA): A comparative hydrological approach. Hydrological Processes. 2001;**15**(8):1441-1457. DOI: 10.1002/hyp.218

[101] Gee GW, Fayer MJ, Rockhold ML, Campbell MD. Variation in recharge at the Hanford site. Northwest Science. 1992;**66**:237-250

[102] Keese K, Scanlon B, Reedy R. Assessing controls on diffuse groundwater recharge using unsaturated flow modeling. Water Resources Research. 2005;**41**:W06010. DOI: 10.1029/2004WR003841

[103] Favreau G, Cappelaere B, Massuel S, Leblanc M, Boucher M, Boulain N, et al. Land clearing, climate variability, and water resources increase in semiarid Southwest Niger: A review. Water Resources Research. 2009;**45**:450-455

[104] Adane ZA, Gates JB. Determining the impacts of experimental forest plantation on groundwater recharge in the Nebraska Sand Hills (USA) using chloride and sulfate. Hydrogeology Journal. 2015;**23**:81-94. DOI: 10.1007/s10040-014-1181-6

[105] Santoni CS, Jobbágy EG, Contreras S. Vadose zone transport in dry forests of Central Argentina: Role of land use. Water Resources Research. 2010;**46**:480-490

[106] Jobbágy EG, Jackson RB. Groundwater use and salinization with grassland afforestation. Global Change Biology. 2004;**10**:1299-1312

[107] Huang T, Pang Z. Estimating groundwater recharge following land-use change using chloride mass balance of soil profiles: A case study at Guyuan

and Xifeng in the loess plateau of China. Hydrogeology Journal. 2011;**19**:177-186. DOI: 10.1007/s10040-010-0643-8

[108] Adane ZA, Nasta P, Gates JB. Links between soil hydrophobicity and groundwater recharge under plantations in a sandy grassland setting, Nebraska Sand Hills USA. Forest Science. 2017;**64**(4):388-401. DOI: 10.5849/FS-2016-137

[109] Owens MK, Lyons RK, Alejandro CL. Rainfall partitioning within semiarid juniper communities: Effects of event size and canopy cover. Hydrological Processes. 2006;**20**:3179-3189. DOI: 10.1002/hyp.6326

[110] Starks PJ, Venuto BC, Dugas WA, Kiniry J. Measurements of canopy interception and transpiration of eastern redcedar grown in open environments. Environment and Natural Resources Research. 2014;**4**:103-122. DOI: 10.5539/enrr.v4n3p103

[111] Scanlon BR, Healy RW, Cook PG. Choosing appropriate techniques for quantifying groundwater recharge. Hydrogeology Journal. 2002;**10**:347. DOI: 10.1007/S10040-002-0200-1

[112] Ali R, McFarlane D, Varma S, Dawes W, Emelyanova I, Hodgson G. Potential climate change impacts on the water balance of regional unconfined aquifer systems in South Western Australia. Hydrology and Earth System Sciences. 2012;**16**:4581-4601. DOI: 10.5194/hess-16-4581-2012

[113] Sanford W. Recharge and groundwater models: An overview. Hydrogeology Journal. 2002;**10**:110-120

[114] Lvovich MI. World Water Resources and their Future. Washington, D.C., USA: American Geophysical Union; 1979

[115] Döll P. Impact of climate change and variability on irrigation requirements: A global perspective. Climatic Change. 2002;**54**:269-293

[116] Alcamo J, Döll P, Henrichs T, Kaspar F, Lehner B, Rösch T, et al. Development and testing of the WaterGAP 2 global model of water use and availability. Hydrological Sciences Journal. 2003;**48**:317-337

[117] Manap MA, Nampak H, Pradhan B, Lee S, Azmin Sulaiman WN, Ramli MF. Application of probabilistic-based frequency ratio model in groundwater potential mapping using remote sensing data and GIS. Arabian Journal of Geosciences. 2014;7:711-724

[118] Verma P, Singh P, Srivastava SK. Development of Spatial Decision-Making for Groundwater Recharge Suitability Assessment by Considering Geoinformatics and Field Data. Arabian Journal of Geosciences. 2020;**13**:306

[119] Tam VT, Nga TTV. Assessment of urbanization impact on groundwater resources in Hanoi, Vietnam, Journal of Environmental Management 2018, 1;227: 107-116. doi: 10.1016/j.jenvman.2018.08.087. Epub 2018 Aug 30.

[120] Siddik MS, Tulip SS, Rahman A, Islam MN, Haghighi AT, Mustafa SMT. The impact of land use and land cover change on groundwater recharge in northwestern Bangladesh. Journal of Environmental Management. 2022;**1**(315):115130. DOI: 10.1016/j.jenvman.2022.115130 Epub 2022 Apr 25

[121] Zomlot Z, Verbeiren B, Huysmans M, Batelaan O. Trajectory analysis of land use and land cover maps to improve spatial–temporal patterns, and impact assessment on groundwater recharge. Journal of Hydrology. 2017;**554**:558-569. DOI: 10.1016/j.jhydrol.2017.09.032

[122] Zomlot Z, Verbeiren B, Huysmans M, Batelaan O. Spatial

distribution of groundwater recharge and base flow: assessment of controlling factors. Journal of Hydrology: Regional Studies. 2015;**4**:349-368. DOI: 10.1016/j.ejrh.2015.07.005

[123] Calder B R, and Smith S M. A time/effort comparison of automatic and manual bathymetric processing in real-time mode, paper presented at U.S. Hydro 2003, U.S. Hydrogr. Soc., Biloxi, Miss., 2003

[124] Fletcher T, Andrieu H, Hamel P. Understanding, management and modelling of urban hydrology and its consequences for receiving waters: A state of the art. Advances in Water Resources. 2013;**51**:261-279

[125] Valle RF, Varandas SGP, Fernandes LS, Pacheco FAL. Groundwater quality in rural watersheds with environmental land use conflicts. Science of the Total Environment. 2014;**493**:812-827

[126] He S, Li P, Wu J, Elumalai V, Adimalla N. Groundwater quality under land use/land cover changes: A temporal study from 2005 to 2015 in Xi'an, Northwest China. Human and Ecological Risk Assessment. 2020;**26**(10):2771-2797

[127] Jia Z, Bian J, Wang Y. Impacts of urban land use on the spatial distribution of groundwater pollution, Harbin City, Northeast China. Journal of Contaminant Hydrology. 2018;**215**:29-38

[128] Tahiru AA, Doke DA, Baatuuwie BN. Effect of land use and land cover changes on water quality in the Nawuni catchment of the white Volta Basin, northern region, Ghana. Applied Water Science. 2020;**10**(8):198

[129] Liu X, Wang X, Zhang L, Fan W, Yang C, Li E, et al. Impact of land use on shallow ground-water quality characteristics associated with human health risks in a typical agricultural area in Central China. Environmental Science and Pollution Research. 2021;**28**(2):1712-1724

[130] Khan R, Jhariya DC. Assessment of land-use and land-cover change and its impact on groundwater quality using remote sensing and GIS techniques in Raipur City, Chhattisgarh, India. Journal of the Geological Society of India. 2018;**92**(1):59-66

[131] He S, Wu J, Wang D, He X. Predictive modeling of groundwater nitrate pollution and evaluating its main impact factors using random forest. Chemosphere. 2022;**290**:133388

[132] Hirschberg KJ. Inventory of known and inferred point sources of groundwater contamination in the Perth basin: Western Australia Department of Mines. Western Australia Geological Survey Record. 1991:1991-1997

[133] Mohamed MM, Elmahdy SI. Remote sensing and information value (IV) model for regional mapping of fluvial channels and to-pographic wetness in the Saudi Arabia. GIScience & Remote Sensing. 2016;**53**:1-22

[134] Nas B, Berktay A. Groundwater quality mapping in urban ground-water using GIS. Environmental Monitoring and Assessment. 2008;**160**(1-4):215-227

[135] Crouvi O, Ben-Dor E, Beyth M, Avigad D, Amit R. Quantitative mapping of arid alluvial fan surfaces using field spectrometer and hyperspectral remote sensing. Remote Sensing of Environment. 2006;**104**:103-117

[136] Singh S, Singh C, Mukherjee S. Impact of land-use and land-cover change on groundwater quality in the lower Shiwalik hills: A remote sensing and GIS based approach.

Open Geosciences. 2010;**2**(2). DOI: 10.2478/v10085-010-0003-x

[137] Zhang F, Huang G, Hou Q, Liu C, Zhang Y, Zhang Q. Groundwater quality in the Pearl River Delta after the rapid expansion of industrialization and urbanization: Distributions, main impact indicators, and driving forces. Journal of Hydrology. 2019;**577**:124004. DOI: 10.1016/j.jhydrol.2019.124004

[138] Alqurashi AF, Kumar L. An assessment of the impact of urbanization and land use changes in the fast-growing cities of Saudi Arabia. Geocarto International. 2019;**34**(1):78-97

[139] Guo Y, Li P, He X, Wang L. Groundwater quality in and around a landfill in Northwest China: Characteristic pollutant identification, health risk assessment, and controlling factor analysis. Expo Health. 2022;**14**:885-901. DOI: 10.1007/s12403-022-00464-6

[140] Li P, Wu J, Qian H, Zhang Y, Yang N, Jing L, et al. Hydrogeochemical characterization of groundwater in and around a wastewater irrigated Forest in the southeastern edge of the Tengger Desert, Northwest China. Expo Health. 2016;**8**(3):331-348. DOI: 10.1007/s12403-016-0193-y

[141] Wakida FT, Lerner DN. Non-agricultural sources of groundwater nitrate: A review and case study. Water Research. 2005;**39**(1):3-16

[142] Townsend MA, Young DP. Assessment of nitrate—Nitrogen distribution in Kansas groundwater, 1990-1998. Natural Resources Research. 2000;**9**(2):125-134

[143] Farooqi A, Masuda H, Siddiqui R, Naseem M. Sources of arsenic and fluoride in highly contaminated soils causing groundwater contamination in Punjab, Pakistan. Archives of Environmental Contamination and Toxicology. 2009;**56**(4):693-706

[144] Vystavna Y, Diadin D, Yakovlev V, Hejzlar J, Vadillo I, Huneau F, et al. Nitrate contamination in a shallow urban aquifer in East Ukraine: Evidence from hydrochemical, stable isotopes of nitrate and land use analysis. Environment and Earth Science. 2017;**76**(13):463

[145] Elmahdy SI, Mohamed MM. Land use/land cover change impact on groundwater quantity and quality: A case study of Ajman emirate, the United Arab Emirates, using remote sensing and GIS. Arabian Journal of Geosciences. 2016;**9**(19):722

[146] Xu F, Li P, Chen W, He S, Li F, Dawei M, et al. Impacts of land use/land cover patterns on groundwater quality in the Guanzhong Basin of Northwest China. Geocarto International. 2022;**37**. DOI: 10.1080/10106049.2022.2115153

[147] Elmahdy S, Mohamed M, Ali T. Land Use/Land Cover Changes Impact on Groundwater Level and Quality in the Northern Part of the United Arab Emirates. Remote Sensing. 2020;**12**(11):1715. DOI: 10.3390/rs12111715

[148] Xu Y, Mo X, Cai Y, Li X. Analysis on groundwater table drawdown by land use and the quest for sustainable water use in the Hebei plain in China. Agricultural Water Management. 2005;**75**(1):38-53

[149] Chen CH, Wang CH, Hsu YJ, Yu SB, Kuo LC. Correlation between groundwater level and altitude variations in land subsidence area of the Choshuichi alluvial Fan, Taiwan. Engineering Geology. 2010;**115**(1):122-131

[150] Dias LCP, Macedo MN, Costa MH, Coe MT, Neill C. Effects of land cover change on evapotranspiration and

streamflow of small catchments in the upper Xingu River basin, Central Brazil. Journal of Hydrology: Regional Studies. 2015;**4**:108-122

[151] Haddeland I, Heinke J, Biemans H, Eisner S, Flörke M, Hanasaki N, et al. Global water resources affected by human interventions and climate change. Proceedings of the National Academy of Sciences USA. 2014;**111**:3251-3256

[152] Brikowski T H, Neku A, Shrestha S D, Smith L.S. Hydrologic control of temporal variability in groundwater arsenic on the Ganges floodplain of Nepal. Water Resources Research. 2014;**518**(Part C):342-353. DOI: 10.1016/j.jhydrol.2013.09.021

[153] Mukherjee A, Bhanja SN, Wada Y. Groundwater depletion causing reduction of baseflow triggering Ganges river summer drying. Scientific Reports. 2018;**8**(1):1-9. DOI: 10.1038/s41598-018-30246-7

[154] Moore WS, Joye SB. Saltwater Intrusion and Submarine Groundwater Discharge: Acceleration of Biogeochemical Reactions in Changing Coastal Aquifers. Frontiers in Earth Science. 2021;**9**:600710. DOI: 10.3389/feart.2021.600710

[155] Prabhakar A, Tiwari H. Land use and land cover effect on groundwater storage. Modeling Earth Systems and Environment. 2015;**1**:45. DOI: 10.1007/s40808-015-0053-y

[156] Schwarzenbach RP, Egli T, Hofstetter TB, Von Gunten U, Wehrli B. Global water pollution and human health. Annual Review of Environment and Resources. 2010;**35**:109-136

[157] Xu P, Feng W, Qian H, Zhang Q. Hydro geochemical characterization and irrigation quality assessment of shallow groundwater in the Central-Western Guanzhong Basin, China. International Journal of Environmental Research and Public Health. 2019;**16**:1492

[158] Izuka SK, Engott JA, Rotzoll K, Bassiouni M, Johnson AG, Miller LD, et al. Volcanic Aquifers of Hawai'i—Hydrogeology, Water Budgets, and Conceptual Models (ver. 2.0, March 2018). Reston, VA, USA: US Geological Survey. 2018. p. 158

[159] Chithra SV, Nair MH, Amarnath A, Anjana NS. Impacts of impervious surfaces on the environment. International Journal of Science and Engineering Invention. 2015;**4**(5):2319-6726

[160] Guzman JA, Moriasi DN, Gowda PH, Steiner JL, Starks PJ, Arnold JG, et al. A model integration framework for linking SWAT and MODFLOW. Environmental Modelling & Software. 2015;**73**:103-116

[161] Gleeson T, Vander Steen J, Sophocleous MA, et al. Groundwater sustainability strategies. Nature Geoscience. 2010;**3**:378-379

[162] van Dijk WM, Densmore AL, Jackson CR, Mackay JD, Joshi SK, Sinha R, et al. Spatial variation of groundwater response to multiple drivers in a depleting alluvial aquifer system, northwestern India. Progress in Physical Geography: Earth and Environment. 2020;**44**(1):94-119. DOI: 10.1177/0309133319871941

[163] Famiglietti JS. The global groundwater crisis. Nature Climate Change. 2014;**4**:945-948

[164] Ferry LT, Mohammed AK, Aslam MA. A Conceptual Database Design for Hydrology Using GIS. In: Proceedings of Asia Pacific Association of Hydrology and Water Resources, Kyoto, Japan. 2003

[165] Asadi SS, Vuppala P, Reddy AM. Remote Sensing and GIS Techniques for

Evaluation of Groundwater Quality in Municipal Corporation of Hyderabad (Zone-V). International Journal of Environmental Research and Public Health. 2007;**4**(1):45-52

[166] Verma A, Thakur B, Katiyar S, Singh D, Rai M. Evalution of ground water quality in Lucknow, Uttar Pradesh using remote sensing and geographic information systems (GIS). International Journal of Water Resources and Environmental Management. 2013;**5**(2):67-76

[167] Hall B, Currell M, Webb J. Using multiple lines of evidence to map groundwater recharge in a rapidly T urbanizing catchment: Implications for future land and water management. Journal of Hydrology. 2020;**580**:124265. DOI: 10.1016/j.jhydrol.2019.124265

[168] Beigi E, Tsai FT-C. Comparative study of climate-change scenarios on groundwater recharge, southwestern Mississippi and southeastern Louisiana USA. Hydrogeology Journal. 2015;**23**:789-806

[169] Ordens MC, Werner AD, Post VEA, Huston JL, Simmons CT, Irvine BM. Groundwater recharge to a sedimentary aquifer in the topographically closed Uley South Basin, South Australia. Hydrogeology Journal. 2012;**20**:61-72

[170] Currell M, Gleeson T, Dahlhaus P. A new assessment framework for transience in hydrogeological systems. Ground Water. 2016;**54**:4-14

[171] Schwartz FW, Sudicky EA, McLaren RG, Park Y-J, Huber M, Apted M. Ambiguous hydraulic heads and 14C activities in transient regional flow. Ground Water. 2010;**48**:366-379

[172] Simpson MJ, Jazaei F, Clement TP. How long does it take for aquifer recharge or aquifer discharge processes to reach steady state? Journal of Hydrology. 2013;**501**:241-248

[173] Peterson TJ, Argent RM, Western AW, Chiew FHS. Multiple stable states in hydrological models: An ecohydrological investigation. Water Resources Research. 2009;**45**:W03406. DOI: 10.1029/2008WR006886

[174] Delin GN, Healy RW, Lorenz DL, Nimmo JR. Comparison of local- to regional-scale estimates of ground-water recharge in Minnesota USA. Journal of Hydrology. 2007;**334**:231-249

[175] Rawling GC, Newton BT. Quantity and location of groundwater recharge in the Sacramento Mountains, south-Central New Mexico (USA), and their relation to the adjacent Roswell Artesian Basin. Hydrogeology Journal. 2016;**24**:757-786

[176] Pan Y, Gong H, Zhou D, Li X, Nakagoshi N. Impact of land use change on groundwater recharge in Guishui River Basin, China. Chinese Geographical Science. 2011;**21**:734-743. DOI: 10.1007/s11769-011-0508-7

[177] Huang TM, Pang ZH, Edmunds WM. Soil profile evolution following land-use change: Implications for groundwater quantity and quality. Hydrological Processes. 2013;**27**(8):1238-1252. DOI: 10.1002/hyp.9302

[178] Crosbie RS et al. Potential climate change effects on groundwater recharge in the High Plains aquifer, USA. Water Resources Research. 2013;**49**(7):3936-3951. DOI: 10.1002/wrcr.20292

[179] Hu W, Wangb YQ, Lic HJ, Huangd MB, Houe MT, Lif Z, et al. Dominant role of climate in determining spatio-temporal distribution of potential groundwater recharge at a regional scale.

Journal of Hydrology. 2019;**578**:124042. DOI: 10.1016/j.jhydrol.2019.124042

[180] Li Z, Zheng FL, Liu WZ. Spatiotemporal characteristics of reference evapotranspiration during 1961-2009 and its projected changes during 2011-2099 on the loess plateau of China. Agricultural and Forest Meteorology. 2012;**154**:147-155. DOI: 10.1016/j.agrformet.2011.10.019

[181] Turkeltaub T, Jia XX, Zhu YJ, Shao MA, Binley A. Recharge and nitrate transport through the deep vadose zone of the loess plateau: A regional-scale model investigation. Water Resources Research. 2018;**54**(7):4332-4346. DOI: 10.1029/2017wr022190

[182] Aeschbach-Hertig W, Gleeson T. Regional strategies for the accelerating global problem of groundwater depletion. Nature Geoscience. 2012;**5**:853-861

[183] Hughes J, Petrone K, Silberstein R. Drought, groundwater storage and stream flow decline in southwestern Australia. Geophysical Research Letters. 2012;**39**(3):L03408. DOI: 10.1029/2011GL050797

[184] Chen J, Wilson C, Tapley B, Scanlon B, Güntner A. Long-term groundwater storage change in Victoria, Australia from satellite gravity and in situ observations. Global and Planetary Change. 2016;**139**:56-65. DOI: 10.1016/j.gloplacha.2016.01.002

[185] Strassberg G, Scanlon B, Chambers D. Evaluation of groundwater storage monitoring with the GRACE satellite: Case study of the High Plains aquifer, Central United States. Water Resources Research. 2009;**45**(5):W05410. DOI: 10.1029/2008WR006892

[186] Bhanja S, Rodell M, Li B, Saha D, Mukherjee A. Spatio-temporal variability of groundwater storage in India. Journal of Hydrology. 2016;**544**:428-437. DOI: 10.1016/j.jhydrol.2016.11.052

[187] Döll P, Flörke M. Global-Scale Estimation of Diffuse Groundwater Recharge: Model Tuning to Local Data for Semi-Arid and Arid Regions and Assessment of Climate Change Impact, Frankfurt Hydrology Paper 03. Frankfurt, Germany: Institute of Physical Geography, Frankfurt University; 2005

[188] Adegoke JO, Pielke RA Sr, Eastman J, Mahmood R, Hubbard KG. Impact of irrigation on midsummer surface fluxes and temperature under dry synoptic conditions: A Fre- gional atmospheric model study of the US High Plains. Monthly Weather Review. 2003;**131**:556-564

[189] Niu GY, Yang ZL, Dickinson RE, Gulden LE, Su H. Development of a simple groundwater model for use in climate models and evaluation with gravity recovery and climate experiment data. Journal of Geophysical Research-Atmospheres. 2007;**112**:D07103. DOI: 10.1029/2006JD007522

[190] Freeze RA, Cherry JA. Groundwater. Englewood Cliffs, NJ: Prentice-Hall; 1979. p. 604

[191] Zhang JE, Felzer BS, Troy TJ. Extreme precipitation drives groundwater re- charge: The northern High Plains aquifer, Central United States, 1950-2010. Hydrological Processes. 2016;**30**(14):2533-2545. DOI: 10.1002/hyp.10809

[192] Shao J, Si B, Jin J. Rooting Depth and Extreme Precipitation Regulate Groundwater Recharge in the Thick Unsaturated Zone: A Case Study. Water. 2019;**11**(6):1232. DOI: 10.3390/w11061232

[193] Lee LJE, Lawrence DSL, Price M. Analysis of water-level response to

rainfall and implications for recharge pathways in the chalk aquifer, SE England. Journal of Hydrology. 2006;**330**(3-4):604-620. DOI: 10.1016/j.jhydrol.2006.04.025

[194] Taylor RG et al. Ground water and climate change. Nature Climate Change. 2013;**3**(4):322-329. DOI: 10.1038/nclimate1744

[195] Jakeman AJ, Barreteau O, Hunt RJ, Rinaudo J-D, Ross A. Integrated Groundwater Management: An Overview of Concepts and Challenges. In: Jakeman AJ, Barreteau O, Hunt RJ, Rinaudo J-D, Ross A, editors. Integrated Groundwater Management. Springer International Publishing; 2016. pp. 3-20. DOI: 10.1007/978-3-319-23576-9_1

[196] Wada Y, Flörke M, Hanasaki N, Eisner S, Fischer G, Tramberend S, et al. Wiberg modeling global water use for the 21st century: The water futures and solutions (WFaS) initiative and its approaches. Geoscientific Model Development. 2016;**9**(1):175-222. DOI: 10.5194/gmd-9-175-2016

[197] Richey AS, Thomas BF, Lo MH, Reager JT, Famiglietti JS, Voss K, et al. Quantifying renewable groundwater stress with GRACE. Water Resources Research. 2015;**51**(7):5217-5237. DOI: 10.1002/2015WR017349

[198] De Waroux YLP, Lambin EF. Monitoring degradation in arid and semi-arid forests and woodlands: The case of the argan woodlands (Morocco). Applied Geography. 2012;**32**:777-786

[199] Geist HJ, Lambin EF. Dynamic causal pattern of desertification. Bioscience. 2004;**54**(9):817-829

[200] Kaptué AT, Prihodko L, Hanan NP. On re-greening and degradation in Sahelian watersheds. Proceedings of the National Academy of Sciences. 2015;**112**:12133-12138

[201] Yoshikawa S, Cho J, Yamada HG, Hanasaki N, Kanae S. An assessment of global net irrigation water requirements from various water supply sources to sustain irrigation: rivers and reservoirs (1960-2050). Hydrology and Earth System Sciences. 2015;**18**:4289-4310

[202] Rulli MC, Saviori A, D'Odorico P. Global land and water grabbing. Proceedings of the National Academy of Sciences. 2013;**110**:892-897

[203] Mather AS. Driving forces. In: Geist HJ, editor. Our earth's Changing Land: An Encyclopedia of Land-Use and Land-Cover Change. Vol. 1. London: (A-K). Greenwood Press, West Port; 2006. pp. 179-185

[204] Li Y, Ni J, Yang Q, Li R. Human impacts on soil erosion identified using land-use changes: A case study from the loess plateau, China. Physical Geography. 2006;**27**(2):109-126

[205] McNeill J, Alves D, Arizpe L, Bykova O, Galvin K, Kelmelis J, et al. Toward a typology and regionalization of land-cover and land-use change: Report of working group B. Changes in Land Use and Land Cover: A Global Perspective. 1994;**4**:55-71

[206] Kariyeva J, van Leeuwen WJ. Phenological dynamics of irrigated and natural drylands in Central Asia before and after the USSR collapse. Agriculture, Ecosystems & Environment. 2012;**162**:77-89

[207] Lave R, Wilson MW, Barron E, Biermann C, Carey M, Duvall C, et al. Critical physical geography. The Canadian Geographer. 2014;**58**(1):1-10

[208] Turner BL, Kasperson RE, Matson PA, McCarthy JJ, Corell RW,

Christensen L, et al. Science and technology for sustainable development special feature: A framework for vulnerability analysis in sustainability science. Proceedings of the National Academy of Sciences. 2003;**100**:8074-8079

[209] Wu J. Landscape sustainability science: Ecosystem services and human well-being in changing landscapes. Landscape Ecology. 2013;**28**:999-1023

[210] Dong SK, Wen L, Yang ZF, Liu SL, Lassoie JP, Zhang XF, et al. Vulnerability of world-wide pastoralism to global changes and interdisciplinary strategies for sustainable pastoralism. Ecology and Society. 2011;**16**:10

[211] Moeck C, Radny D, Popp A, Brennwald M, Stoll S, Auckenthaler A, et al. Characterization of a managed aquifer recharge system using multiple tracers. Science of the Total Environment. 2017;**609**:701-714

[212] Götzinger J, Barthel R, Jagelke J, Bárdossy A. The role of groundwater recharge and baseflow in integrated models. IAHS-AISH Publication. 2008;**321**:103-109. DOI: 10.13140/2.1.2192.8960

[213] Yifru BA, Chung IM, Kim MG, Chang SW. Assessment of groundwater recharge in agro-urban watersheds using integrated SWAT-MODFLOW model. Sustainability (Switzerland). 2020;**12**(16):1-18. DOI: 10.3390/su12166593

[214] Shuler C, Brewington L, El-Kadi AI. A participatory approach to assessing groundwater recharge under future climate and land-cover scenarios, Tutuila, American Samoa. Journal of Hydrology: Regional Studies. 2021;**34**(September 2020):100785. DOI: 10.1016/j.ejrh.2021.100785

Chapter 2

Assessment of Heavy Metals Contamination in Groundwater and Its Implications for Public Health Education: A Case Study of an Industrial Area in Southwestern Nigeria

Titus Oladapo Okareh, Adewale Allen Sokan-Adeaga, Tosin Akin-Brandom, Micheal Ayodeji Sokan-Adeaga and Eniola Deborah Sokan-Adeaga

Abstract

Heavy metals' presence in groundwater has garnered a lot of attention recently due to their impact on ecosystem and human health. Thus, this chapter was designed to assess the effects of heavy metals contamination in groundwater and its implication for public health education in selected communities located in an industrial area in Ogun state, southwestern Nigeria. Fifty groundwater sources were identified using a handheld global positioning system and analyzed for physico-chemical and heavy metals properties. Four-hundred participants were selected and interviewed using pre-tested semi-structured questionnaire. The results indicated that there were high quantities of heavy metals in the groundwater that were above the allowable limit set by national and international regulations. A larger percentage of the respondents' drinks water from groundwater supply. The majority of those surveyed were poorly informed on the effects of heavy metal contamination. The following ailments were experienced by the respondents in the last 6 months: frequent watery stool, difficulty in breathing, and skin infection. Few of the respondents reported the following occurrences in their household in the last 1 year: still birth, stunted growth in child, and death due to cancer. Therefore, there is a requirement for immediate public health education and health promotion activities among the local populace.

Keywords: groundwater, heavy metals toxicity, public health, environmental monitoring, community education, industrial activities, physicochemical properties

1. Introduction

Human life depends critically on water. For the survival of all living things, it is absolutely essential. The best source of freshwater with the smallest amount of salts for human consumption is groundwater [1]. Unchecked population explosion, uncontrolled urbanization, and inappropriate disposal of solid and liquid wastes all contribute to the intrusion of hazardous materials into underground water supplies [2]. The two main causes of groundwater pollution are the unregulated discharge of industrial wastes and the use of chemical fertilizers in agriculture [3]. The increased use of water by people, particularly as a receptacle to dispose of human waste, is another important factor contributing to groundwater pollution. The impacts of additional organic matter and pathogens are of public health significance [4, 5]. Additionally, because of the soil's porosity, sewage, and leachate from wastes are more likely to enter subsurface water bodies, which are an important supply of water for many towns [6]. Water-borne infections have created a significant epidemic of sickness as a result of fecal waste contamination of drinking water [7, 8].

One of the main pollutants in groundwater sources is heavy metals [9]. Some of these heavy metals are necessary for an organism's growth, development, and health, whereas others are not since they are irreplaceable and the majority of them are harmful to living things [10]. But the concentration of heavy metals in the environment determines how harmful they are. Heavy metals leach into groundwater and soil solution as ambient concentrations rise and soils lose their capacity to retain them. These dangerous heavy metals can consequently accumulate in living tissues and concentrate at the top of the food chain [11]. Effects of heavy metals bioaccumulation at biochemical levels often include the replacement of required ions, harm to plasma membranes, interactions with sulfhydryl (-SH) groups, reactions with phosphate ions, and competition for binding sites with important metabolites [12, 13].

In Nigeria, boreholes and dug wells provide a significant portion of the country's drinking water supply. Since groundwater is often the primary supply of drinking water in rural and some urban regions, a sizable population is at danger of ingesting contaminated water. Groundwater quality is influenced by a variety of factors, including aquifer lithology, groundwater velocity, the quality of recharge fluids, interactions with other types of water or aquifers, human activities, and the environment [14]. Environmental monitoring and impact assessment programs have attracted a surge of interest from planners and environmentalists concerned with the environmental repercussions of companies. Scott [15] asserts that there has been a tendency to neglect how industries in the developing countries affect the environment. Although the growth of these firms is regarded to be a way to increase employment and earnings, there is not enough information on their environmental impact and sustainability to support decision-making.

The industrial sectors in Shagamu and Otta are thought to be Ogun state's fastest-growing areas, located in the southwestern part of Nigeria. Groundwater is largely one of the geological resources that have been negatively impacted by this expansion. When surface water supplies are no longer sufficient to satisfy the needs of communities, groundwater is the only other source of high-quality water. However, it has been shown that some regions of the industrial belts have a number of problems with groundwater contamination [14]. The conurbation of Shagamu and Otta has a sizable number of diverse industries, all of which have been developed. As a result, these industries have been dangerously degrading the quality of groundwater by releasing pollutants into the ambience at ever-increasing amounts. Hence the main objectives of this chapter are to: (1) characterize the physicochemical and heavy metal properties of groundwater collected from the selected communities in line with national and international permissible limits, (2) assess the knowledge level of households on toxicities associated with heavy metals contamination of groundwater, and (3) assess the health status of respondents' households in the last six (6) months.

2. Methodology

In this section, we shall discussed the step by step methodology employed in the assessment of heavy metals contamination in groundwater; and the survey on the knowledge level and health implication of these contaminations among residents in the concerned areas. We shall begin with the study design and other baseline information; followed by the environmental sampling and laboratory analytical procedure; and end this section with cross-sectional descriptive survey.

2.1 Study design and baseline information

In order to accomplish the study's stated aims, two (2) designs were used: a laboratory investigation and a cross-sectional descriptive survey. The study was conducted in Ogun State's Shagamu Local Government Area (SLGA), which is in Nigeria's southwest. On September 23, 1991, the former Remo Local Government was divided into the SLGA. Its borders are Odogbolu Local Government, Lagos State, Ikenne Local Government, and Obafemi Owode Local Government in that order. According to the 2006 census, it has a land area of 605.6 sq. km and a population of 255,885. According to the 2006 National Population Commission, the expected population in 2022 will be 435,200, growing at a rate of 3.4% year. The area is divided into fifteen (15) wards for administrative and political convenience, namely: Oko Epe and Itunla 1—I; Oko Epe and Itunla—II; Aiyegbami/Ijoku—III; Sabo 1—IV; Sabo 11—V; Itunsoku/Oyebajo—VI; Ijagba—VII; Latawa—VIII; Ode-lemo—IX; Ogijo/Ikosi—X; Surulere—XI; Isote—XII; Simawa—XIII; Agbowa—XIV; and Ibido/Itn Alara—XV. Due to the existence of several industries, indiscriminate waste disposal methods, and high industrial activity practices, Sabo1, Sabo 11, and Ogijo/Ikosi were chosen study sites [16].

2.2 Laboratory study

2.2.1 Environmental sampling (coordinates acquisition)

Using a GPS Garmin 60 on-site at the groundwater source location, samples were located. The GPS was activated and given authorization to use satellite signals for navigation. Following the receipt of full signals, the GPS device's coordinates for the groundwater source were registered, downloaded, and inserted into the database to produce a map that shows the positions of the samples throughout the study area. In order to maintain track of each unique location, the way points were kept to correlate with the names of each groundwater supply site.

2.2.2 Sample collection and transport

Fifty groundwater samples were purposefully collected for laboratory analysis from the chosen study sites (Sabo1, Sabo 11, and Ogijo/Ikosi). Sterilized 500 ml bottles were used to collect the water samples. A portable GPS unit was used to find the groundwater locations (GPS). Using a multifunctional digital pH meter, the pH, temperature, TDS, and conductivity of the groundwater samples were promptly determined. Concentrated hydrogen trioxonitrite (IV) (HNO_3) was used to preserve water samples before they were sent directly to the lab for testing.

2.2.3 Physicochemical analysis of sample

The following physicochemical properties were assessed for in the collected water samples:

i. pH: Using a calibrated multifunctional conductivity, total dissolved solids (TDS), temperature, and pH meter, the pH values of the water samples were obtained (Hanna HI 9811-5 model). The pH meter probe was inserted into the sample containers. After the LCD display had stabilized, the pH readings were then taken.

ii. Conductivity: The capacity of a substance to transport heat, electricity, or sound is referred to as its conductivity. Using a conductivity meter (Hanna HI 9811-5 model), the conductivity of the 50 water samples was determined. Once a constant number was obtained, the results were read.

iii. Temperature: Water temperature is the degree Celsius reading of the water sample at the moment it was taken. The time of day and the current weather conditions both play a role. The rate of chemical reaction in the water accelerates as temperature rises. The multipurpose instrument (Hanna HI 9811-5 model) was used to determine the temperature of the samples. A 500 ml sterilized bottle was placed into the already switched-on meter, with the electrode tip touching the water. The outcome was read once a reliable reading had been acquired. The identical procedure was used to each of the 50 water samples.

iv. Total dissolved solids (TDS): A measure of dissolved solids in an aqueous solution is called TDS. "Dispersed solids" refers to any minerals, salts, cations, or anions that have dissolved in water. TDS is made up of certain trace amounts of organic material as well as dissolved inorganic salts like calcium, magnesium, potassium, sodium, bicarbonates, chlorides, and sulfates. TDS is a test that evaluates the overall water quality. Since the TDS content is more of an esthetic than a health concern, it is regulated as a secondary drinking water standard. Though it is not linear, there can be a correlation between a TDS content and water conductivity. The TDS of the 50 water samples was determined using a TDS/conductivity meter (Hanna HI 9811-5 model).

2.2.4 Heavy metals determination of samples

The Association of Analytical Chemists (AOAC) methodologies were used to determine the concentration of heavy metals in groundwater samples as outlined below:

2.2.4.1 Digestion of samples

After cleaning and drying the equipment, 10% aqua-regia (HCl, HNO_3 ratio: 3:1) was used to rinse it. In a clean, 250 ml conical flask, 5 ml of pure nitric acid (HNO_3), and 100 ml of the preserved sample were added. The mixture was cooked on a hot plate inside the fume cupboard until it was almost dry. A further addition of 10 ml of distilled deionized water and 1 ml of hydrogen peroxide was added. The digesting vessel was covered with a crucible and heated for a further 5 minutes. The digest was quantitatively transferred to a 50 ml volumetric flask after cooling, with the volume being made up with distilled deionized water. It was then labeled and analyzed for the parameter(s) of interest. The other samples and the blank were treated in the same way.

2.2.4.2 Determination of heavy metals concentrations (As, Cd, Cr, and Pb)

After digestion, the samples were tested for arsenic (As) at 193.7 nm, cadmium (Cd) at 228.8 nm, chromium (Cr) at 357.9 nm, and lead (Pb) at 283.3 nm using the Perking 3300 AAS. A Perkin Elmer MHS-10 hydride generator was utilized to measure the system's As. The analytical conditions for the standard and blank assays for each metal were the same. The AAS readout, sample volume acquired for analysis, and extract volume were all used to compute the metal content in each sample.

$$\text{Metal(mg/L)} = \frac{(\text{The result} - \text{Blank}) \times \text{Vol.of extract}}{\text{Vol.of Sample taken}} \quad (1)$$

Note: Both the physicochemical and heavy metals analyses results were compared with permissible values/limits specified by both International and local bodies vis World Health Organisation (WHO) [17], and Nigerian Standard for Drinking Water Quality (NSDWQ) [18].

2.3 Household survey (cross-sectional study)

2.3.1 Sample size determination

The number of households needed for the survey was determined using Leishie Kish formula.

$$N = \frac{Z^2P(1-P)}{d^2} \quad (2)$$

where
N = Sample size
Z = The value of the normal variant, confidence level of 1.96 for the 95% confidence interval.
P = The expected prevalence rate (in this case being a new study, 50% prevalence was considered (0.5)).
d = The highest acceptable absolute precision in % (±5%) = 0.05 error in the estimate substituting the formula, given:

$$N = \frac{(1.96)^2 \times 0.5(1-0.5)}{(0.5)^2} \tag{3}$$

$$= \frac{3.84 \times 0.5(0.5)}{0.0025}$$

$$= 384 \text{ individuals}$$

This was rounded up to 400.

2.3.2 Sampling technique

A three stage multiple sampling process was used. From the local government area, 15 wards were identified. Three of the 15 wards were specifically chosen because of their heavy industrial activity and indiscriminate waste disposal. Seven communities were purposively selected from the three wards based on their high levels of industrial activity and careless waste disposal, both of which are risk factors for heavy metal poisoning of the environment. The systematic sampling method was used to choose households from among the seven communities. This entails picking a family at random, sampling every subsequent nth = 3rd house, that is, first, fourth, and seventh.

2.3.3 Instrument for data collection

The authors devised a structured, interviewer-administered questionnaire that was pretested for data collection. The survey was divided into four components, including:

Section A: socio-demographic information of the respondents. Section B: general information on water supply. Section C: assessed respondents' knowledge level on heavy metals problems. Knowledge was scored based on seven (7) items with each correct answer attracting 1 mark. The total mark was 7, while the lowest score was 0. Respondents that scored between 0 and 2 were grouped under poor knowledge; those that scored between 3 and 5 were grouped under fair knowledge while those that score between 6 and 7 were regarded as having good knowledge. Section D: respondent's information on health status. The questionnaires contained both open and close ended questions.

2.3.4 Validity and reliability of the instrument

The test-retest method was employed to evaluate reliability. This strategy entailed administering the same instrument to the same subjects on various occasions while operating under comparable assumptions. The results of multiple tests were compared. The questionnaire was distributed twice over the course of 2 weeks to 40 households representing 10% of the sample size in Odogunyan, Lagos State, which is about 1 km from the study area and is also susceptible to heavy metal contamination due to industrial activities and careless trash disposal. The questionnaire's consistency was checked using the pre-testing. The reliability coefficient was calculated to assess the instrument's dependability for the study and check for internal consistency of answer. The pre-reliability test's value was 0.76, which demonstrated the validity of the questionnaire.

2.3.5 Data collection procedure

Face-to-face interviews were used to gather the data over the course of twelve (12) weeks.

2.3.6 Data management and analysis

After data collection, the questionnaires underwent a careful completion check. The data was manually entered and analyzed using IBM Statistical Product and Service Solutions version 23. The examined data were represented using descriptive statistics (mean and standard deviation), frequency tables, percentages, and charts.

2.3.7 Ethical consideration

The Babcock University Health Research Ethics Committee (BUHREC) accepted the protocol and gave the study its ethical approval. The study was conducted in compliance with the Helsinki Declaration. A letter of introduction to the Chairman of the Sagamu Local Government Area was acquired from the Department of Public Health prior to the start of the study. The communities were allowed admission after receiving a letter of approval from the local authorities. Additionally, before being enrolled in the study, all willing individuals verbally consented after being informed about the research. Those who agreed were thumb printed or signed before being questioned (for illiterate participants). The respondents were given the promise that the information they provided would be utilized only for research, and the researchers strictly protected the privacy and confidentiality rights of the study participants.

3. Results and discussion

This section deals with findings and discussions of the results obtained from the lab work and field survey. It presents the physicochemical properties and heavy metals analysis results of the water samples collected from the various sampling sites. It also highlights the findings from the socio-demographic characteristics, general information on water supply, and knowledge of respondents on the health impacts of heavy metals contamination. The results were presented using frequency tables, charts, line graphs, mean, and standard deviation.

3.1 Physicochemical properties of water samples

The findings of the physicochemical analyses of the groundwater samples are shown in **Table 1** below, together with the permitted limitations for each parameter set by the World Health Organization (WHO) and Nigerian Standard for Drinking Water Quality (NSDWQ). The samples' pH ranged between 3.8 and 7.9, with a mean value of 4.79 ± 1.6. The temperature had a mean of 30.09°C and a range of 27.7 to 34.5°C. A mean value of 81.25 ± 5.8 μ^2/cm was obtained for the conductivity value, which ranged from 010 to 750 μ^2/cm. With a mean value of 503.9 mg/L, the total dissolved solids (TDS) value ranged from 001 to 3360 mg/L.

The pH of the sampled water was outside the range (6.5–8.5) specified by the WHO/NSDWQ. This result is an indication that the groundwater in the selected communities is acidic, which potent ill to the health and well-being of the populace

Parameter	Range	Means/SD	WHO 2011/NSDWQ 2008
pH	3.8–7.9	4.794 ± 1.6	6.5–8.5
Temperature (°C)	27.7–34.5	30.093 ± 3.6	Ambient
Conductivity (μ^2/cm)	010–750	81.25 ± 5.8	500
Total dissolved solids (mg/L)	0010–0360	50 ± 3.9	500–600

Table 1.
Physicochemical properties of groundwater samples.

who depend on it as their source of drinking water. This acidity may be attributed to the high industrial activities and emission of noxious gases such as CO_2, SO_2, and NO_3 which formed carbonic, sulfuric, and nitrous acid in the air; and later precipitates as weak acid during rainfall and infiltrate into the groundwater [14, 19]. Acidosis may result from consuming such samples of water on a regular basis because the pH is too low (acidic pH) [20]. Similar to this, acidic water has been linked to mucous membrane cell destruction and skin and eye discomfort [21]. This conforms to the findings of Longe and Balogun [22], who reported a 5.30–7.07 range in Lagos, Nigeria groundwater. The value's proximity may be the result of geological comparison. The groundwater mean temperature value observed in this study was comparable to those reported by Ojekunle et al. [14] in Sango-Otta, Okareh et al. [19] in Shagamu, and by Rahim and Hussain [23] in an industrial area of India. All the authors found groundwater temperatures to range from 28 to 30°C.

Ions are detected in water by measuring its electrical conductivity (EC), a measure of total dissolved solids. Shagamu's EC values ranged from 010 to 750 μ^2/cm. This may be connected to the local lithology and predominate anthropogenic activities. Furthermore, this outcome is comparable to earlier work by Refs. [14, 19, 24]. The term "total dissolved solid" (TDS) refers to inorganic salts. The measured TDS values were within the limits of Standard Organisation of Nigeria (SON) [18] and World Health Organisation (WHO) [17] standards, which call for 500 and 600 mg/L, respectively. The range of the TDS values obtained in groundwater for the current investigation is less than 1000 mg/L, as shown in **Table 1**, and can therefore be categorized as freshwater [25, 26]. The outcome is consistent with related research by Refs. [19, 27].

3.2 Heavy metals concentrations in groundwater

The groundwater of the study area was analyzed for the presence of four heavy metals: As, Cd, Cr, and Pb. The varied heavy metal concentrations found in fifty (50) samples of groundwater are shown in **Tables 2–5** and **Figure 1**. Arsenic (As) concentrations ranged from 0.0001 to 0.0089 mg/L, with a mean value of 0.002–0.002 mg/L. The levels of As in every source of water that was sampled are all below the WHO/NSDWQ-approved permitted limits. The minimum and maximum concentrations that were found were <0.0001 and 0.0089, respectively. With a mean concentration of 0.080.11, the cadmium (Cd) content ranges from 0.01 to 0.43 mg/L. Only 52% of the water sources that were analyzed had Cd concentrations that were within the permissible range outlined by WHO/NSDWQ. The minimum and maximum concentrations that were found were <0.01 and 0.43 mg/L, respectively. With a mean concentration of 0.03–0.05 mg/L, the value for the chromium (Cr)

Parameters	Borehole water	Well water	Both
Number of total samples	41	9	50
Number of arsenic detected within WHO/NSDWQ limit (MCL)	43	7	50
Percentage of arsenic detected within WHO/NSDWQ limit (MCL)	86%	14%	100%
Number of arsenic detected above WHO/NSDWQ limit (MCL)	0	0	0
Percentage of arsenic detected above WHO/NSDWQ limit (MCL)	0	0	0
Minimum concentration detected (mg/L)	<0.0001	0.0001	<0.0001
Maximum concentration detected (mg/L)	0.0089	0.00061	0.0089
WHO/NSDWQ maximum contaminant level (MCL) (mg/L)	0.01	0.01	0.01
Mean	0.002	0.002	0.002

Table 2.
Levels of arsenic concentration in groundwater samples.

Parameters	Borehole water	Well water	Both
Number of cadmium detected within WHO/NSDWQ limit (MCL)	19	7	26
Percentage of cadmium detected within WHO/NSDWQ limit (MCL)	38%	14%	52%
Number of cadmium detected above WHO/NSDWQ limit (MCL)	23	1	24
Percentage of cadmium detected above WHO/NSDWQ limit (MCL)	46%	2%	48%
Minimum concentration detected (mg/L)	<0.01	0.12	< 0.01
Maximum concentration detected (mg/L)	0.43	0.12	0.43
WHO/NSDWQ MCL (mg/L)	0.003	0.003	0.003
Mean	0.08	0.08	0.08

Table 3.
Levels of cadmium concentration in groundwater samples.

Parameters	Borehole water	Well water	Both
Number of chromium detected within WHO/NSDWQ limit (MCL)	32	7	39
Percentage of chromium detected within WHO/NSDWQ limit (MCL)	64%	14%	78%
Number of chromium detected above WHO/NSDWQ limit (MCL)	11	—	11
Percentage of chromium detected above WHO/NSDWQ limit (MCL)	22%	—	22%
Minimum concentration detected (mg/L)	<0.001	—	<0.001
Maximum concentration detected (mg/L)	0.215	—	0.215

Parameters	Borehole water	Well water	Both
WHO/NSDWQ MCL (mg/L)	0.05	0.05	0.05
Mean	0.03	0.03	0.03

Table 4.
Level of chromium concentration in groundwater samples.

Parameters	Borehole water	Well water	Both
Number of lead detected within WHO/NSDWQ limit (MCL)	6	—	—
Percentage of lead detected within WHO/NSDWQ limit (MCL)	12%	—	12%
Number of lead detected above WHO/NSDWQ limit (MCL)	37	7	44
Percentage of lead detected above WHO/NSDWQ limit (MCL)	74%	14%	88%
Minimum concentration detected (mg/L)	<0.01	0.13	0.01
Maximum concentration detected (mg/L)	3.26	3.12	3.26
WHO/NSDWQ MCL (mg/L)	0.01	0.01	0.01
Mean	0.51	0.51	0.51

Table 5.
Level of lead concentration in groundwater samples.

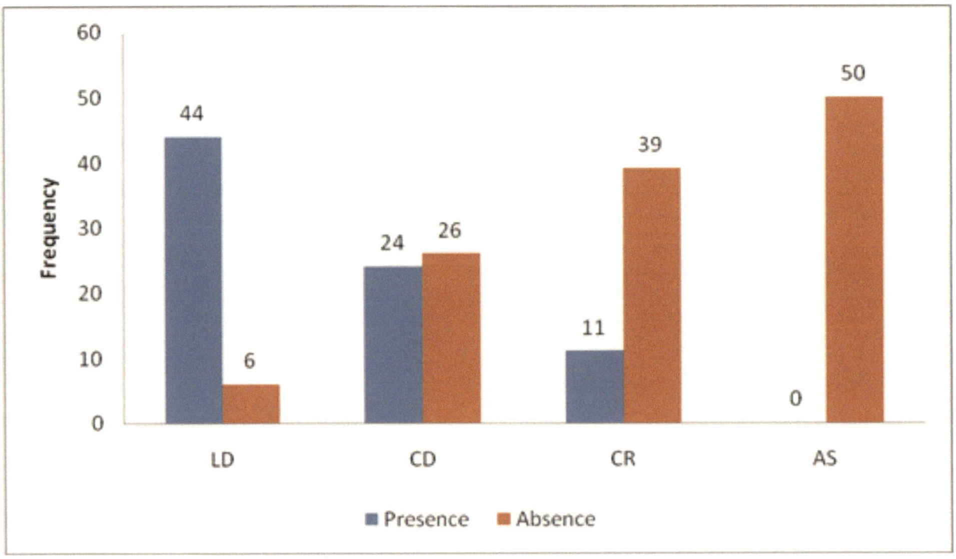

Figure 1.
Levels of concentrations of heavy metals (Pb, Cd, Cr, and As) in groundwater sample.

determination ranges from 0.001 to 0.215 mg/L. The WHO/NSDWQ maximum permissible level for Cr concentration is met by more than a third (78%) of the analyzed water sources. The minimum and maximum concentrations that were found were

0.001 and 0.215 mg/L, respectively. The concentration of lead (Pb) varied from 0.01 to 3.26 mg/L, with a mean of 0.51–0.71 mg/L. A few (12%) of the water sampled has Pb concentration within the WHO/NSDWQ permissible limit. The concentration ranges from 0.01 mg/L at the lowest level to 3.26 mg/L at the highest level.

Cadmium comes from two different sources: byproducts of the zinc refining process and naturally occurring ores in rocks and soils [28]. By coming into contact with soil that had been contaminated by discharges from the mining, paint, electroplating, petrochemical, plastics, and fertilizer sectors, groundwater became contaminated with cadmium by leaching [29]. Epidemiological research has shown that chronic exposure to Cd may cause kidney damage, lung cancer, high blood pressure, and bone abnormalities (osteoporosis and osteomalacia) [13, 30]. This is true even though the Cd levels in more than half of the sampled groundwater sources were below the permitted limit (0.003 mg/L) established by the WHO and NSDWQ. Chromium is a naturally occurring element found in volcanic emissions, rocks, soil, plants, and animals. The main types are trivalent (chromium 3) and hexavalent (chromium 6) and can be found in drinking water. The main causes of Cr presence in the groundwater samples could have been natural deposit erosion and coating removal from water pipelines [13, 30]. Despite the fact that Cr concentrations were below the 0.05 mg/L WHO and NSDWQ acceptable range, the health effects of excessive chromium exposure include hepatic and renal impairment while chromate dust is carcinogenic [31, 32].

Due to its deadly and lethal character even at extremely low concentrations, lead is the most significant heavy metal [33]. It can accumulate in body tissue, putting people's health in peril. The samples examined at various locations contained lead concentrations ranging from 0.01 to 3.26 mg/L, respectively. The majority of the samples revealed lead contents over the 0.01 mg/L permissible level established by the WHO and NSDWQ. The Environmental Protection Agency's (EPA) permitted threshold is 0 mg/L due to its toxicity. (1) The nearby paint industry's discharge of lead-rich waste effluents that were deposited in the soil and later made their way into underground water via leaching and (2) the dissolution of industrial heavy plant aerosols and dusts into the soil by heavy rain may be the causes of the high lead concentrations found in some of the sampled groundwater [30, 34]. Hypertension, disturbance of vitamin D and calcium metabolism, impairment of fetal and young children's brain development, harm to human tissues and organs, and many other difficulties can arise from excessive lead levels in water [13, 30, 35]. Arsenic is present in sulphide complexes such as realgar (As_2S_2), orpiment (As_2S_3), and iron pyrites, to name a few [36]. It is acknowledged as a poison and human carcinogen. Arsenic is included as the most important contaminant at superfund sites on the ATSDR/Environmental Protection Agency (EPA) priority list [37]. Arsenate (+5) predominates when it is hydrated, though arsenite (+3) predominates when anaerobic conditions are present. Although the concentration in groundwater can be significantly greater, its usual concentration in natural streams is less than 1–2 mg/L [13]. Despite the fact that all of the groundwater samples were within the WHO and NSDWQ permissible limits for arsenic concentrations, there is still a chance that arsenic will bioaccumulate in biological systems. The World Health Organization [38] states that melanosis, an abnormal black-brown skin pigmentation, and keratosis, a hardening of the palms and soles, are the first noticeable effects of exposure to low levels of arsenic in drinking water. Keratosis can thicken further (hyperkeratosis), which can lead to skin cancer.

3.3 Socio-demographic characteristics of respondents

The mean age of respondents were 33.7 ± 3.4 years with majority falling within the age group of 30–39 years. Nearly a third-quarter 295(73.8%) of the respondents was males and majority 299(74.8%) has secondary school education. The predominant ethnic group was Yoruba 335(83.8%). Majority 285(71.2%) of the respondents had stay in their present residence for 4 years and above as depicted in **Table 6**. The high proportion of youth in the survey is suggestive of the demographic distribution of a typical urban settlement in sub-Saharan Africa which serve as a commercial and employment hub for young people. Also the high percentage of Yoruba ethnic group is anticipated since the study was conducted in the southwestern part of Nigeria. These findings are in consonance with the reports of previous authors [6, 39]. The demographic of the survey communities' also revealed high proportion of semi-literate and artisans/traders. This highlights the poor socio-economic status of residents in the selected communities [39].

Variables	Frequency	Percentage
Age group (year)		
10–19	0	0.00
20–29	48	12.0
30–39	186	46.5
40–49	123	30.8
≥50	43	10.7
Total	400	100.00
Gender		
Male	295	73.8
Female	105	26.2
Total	400	100.0
Level of education		
None	14	3.5
Primary	58	14.5
Secondary	299	74.8
Tertiary	29	7.2
Total	400	100.0
Types of occupation		
Civil servant	81	20.2
Artisan	125	31.2
Farming	7	1.8
Others (trader/house wife)	187	46.8
Total	400	100.0
Ethnicity		
Yoruba	335	83.8

Variables	Frequency	Percentage
Igbo	44	11.0
Hausa	21	5.2
Total	400	100.0
Position of respondents in the household		
Landlord	168	42.0
Landlady	84	21.0
Tenant	148	37.0
Total	Total	100.0
Duration of living in present residence (in years)		
2–3	115	28.8
4–5	128	32.0
6–7	102	25.5
8–9	39	9.7
≥10	16	4.0
Total	400	100.0

Mean age of respondents = 33.7 ± 3.4 years.

Table 6.
Socio-demographic characteristics of respondents.

3.4 Information on water supply

Table 7 presents the information on water supply of the respondents. Majority 357 (89.2%) of the respondents reported borehole has the source of water supply in their household. More than a third-quarter 316(79.0%) of the respondents drinks water from groundwater supply while the remaining 84(21.0%) drinks sachet water. Two hundred and fifty eight (81.7%) of the respondents reported that they do not treat their groundwater before drinking while the remaining reported using the following form of treatments: sedimentation; 29(9.2%), boiling; 14(4.4%), coagulation; 7 (2.2%), and chlorination; 8(2.5%). The reasons stated by some 258(81.7%) of the respondents for not treating the groundwater before drinking include: groundwater is less harmful; 141(54.7%), no prior knowledge about the treatment; 29(11.2%), it is time wastage; 42(16.3%), drinking it raw from childhood without any negative consequences; 46(17.8%).

Groundwater contributes to both the piped and non-piped domestic water supplies that are found in towns and cities throughout sub-Saharan Africa [40, 41]. Individual choices for household water supply are influenced by issues with accessibility, affordability, dependability, or convenience [42]. The Sustainable Development Goal, objective 6.1, takes availability, accessibility, and safety into consideration. In this study, majority of the households depend on groundwater source for their domestic water supply. The relative closeness of the water supply to the home, which can be acquired at relatively low recurrent cost and, in some circumstances, with cheap upfront inputs, is one of the benefits of self-supply

Variables	Frequency	Percentage
Source of water supply in household		
Borehole	357	89.2
Hand-dug well	43	10.8
Total	400	100.0
Do you drink water from groundwater source?		
Yes	316	79.0
No	84	21.0
Total	100	100.0
Source of drinking water		
Sachet	84	21.0
Borehole	306	76.5
Hand-dug well	10	2.5
Total	400	100.0
When was your groundwater source dug?		
2–3	56	14.0
4–5	106	26.5
6–7	23	5.7
8–9	16	4.0
≥10	8	2.0
No idea	191	47.8
Total	400.0	100.0
Estimated depth of groundwater source		
70–89	56	14.0
90–109	58	14.5
110–129	39	9.8
130–149	0	0.0
≥150	14	3.5
No idea	233	58.2
Total	400	100.0
Treatment technique employed by household to make groundwater source safe		
No treatment technique	258	81.7
Sedimentation	29	9.2
Boiling	14	4.4
Coagulation	7	2.2
Chlorination	8	2.5
Total	316	100.0
If yes to treatment technique, how often?		
Once in a while	9	15.5

Variables	Frequency	Percentage
Once in 3 month	10	17.2
Once in a year	12	20.7
I cannot say	27	46.6
Total	58	100.0
If no treatment technique, state reasons		
Borehole/well water is less harm.	141	54.7
No prior knowledge about it.	29	11.2
Time wastage.	42	16.3
I have been drinking it from childhood and it has no inimical effect on me.	46	17.8
Total	258	100

Table 7.
Water supply information.

using groundwater sources [42, 43]. Such a supply is available to many houses and can enable a relatively quick response to increasing needs where hydrogeology is advantageous and drillers can provide their services. Aquifers of groundwater, in general, offer long-term storage and can serve as a buffer during dry spells [44–46]. Furthermore, the majority of homes do not treat their water in any way before using it for drinking. They may have been content with the water's apparent clarity since they lacked awareness about the toxicity of heavy metals and other toxins found in groundwater. This supported the findings of Abolanle-Azeez et al. [47], who claimed that most households in a small group of Ogun State communities did not treat their water.

3.5 Knowledge of heavy metals contamination among respondents

From **Table 8**, barely more than one-fifth, 108(27.0%) of the respondents have erudition that groundwater source can be contaminated by heavy metals pollution. Some of the health problems reported by respondents to be caused by drinking water contaminated with heavy metals include: mouth odor; 15(3.8%), poisoning; 79

Variables	Frequency	Percentage
Do you know groundwater source may be contaminated by heavy metals?		
Yes	108	27.0
No	160	40.0
Do not know	132	33.0
Total	400	100.0
Drinking water contaminated with heavy metals may cause any of these?		
Mouth odor	15	3.8
Poisoning	79	19.7
Malaria	63	15.7

Variables	Frequency	Percentage
I do not know	243	60.8
Total	400	100.0
Arsenic exposure in drinking water can cause the following?		
Stomach ulcer	7	1.7
Typhoid	106	26.6
Dermal problem	7	1.7
I do not know	280	70.0
Total	400	100.0
Cadmium exposure in drinking water can cause the following?		
Typhoid	77	19.2
Kidney problem	14	3.5
I do not know	309	77.3
Total	400	100.0
Chromium exposure in drinking water can cause the following?		
Typhoid	78	19.5
Liver problem	8	2.0
Malaria	21	5.3
I do not know	293	73.2
Total	400	100.0
Lead exposure in drinking water can cause the following?		
Typhoid	91	22.7
Malaria	14	3.5
I do not know	295	73.8
Total	400	100.0
Chronic exposure to heavy metals in drinking water may cause any of these?		
Typhoid	121	30.2
Cancer	67	16.8
Frequent watery stooling	77	19.2
I do not know	135	33.8
Total	400	100.0

Table 8.
Knowledge of respondents on heavy metals contamination.

(19.7%), and malaria; 63(15.7%). Respondents believed that arsenic contamination in drinking water can cause the following: typhoid; 106(26.6%), stomach ulcer; 7(1.7%), and dermal problem; 7(1.7%). The respondents reported the following as the health problems associated with cadmium exposure in drinking water: typhoid; 77(19.2%), and kidney problem; 14(3.5%), while chromium exposure in drinking water is believed to cause typhoid; 78(19.5%), liver problem; 8(2.0%); and malaria; 21(5.3%).

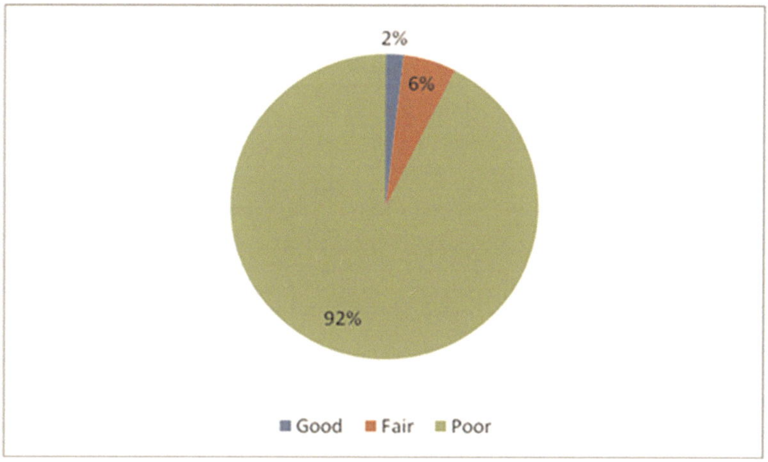

Figure 2.
Respondents knowledge level.

Ninety-one (22.7%) and 14 (3.5%) of the respondents reported that lead exposure in drinking water can cause typhoid and malaria, respectively. The respondents associated the following to chronic exposure to heavy metals contaminations in drinking water: typhoid; 121(30.2%), cancer; 67(16.8%), and frequent watery stooling; 77 (19.2%). From **Figure 2**, 368(92.0%) of the respondents had poor knowledge of the consequences of heavy metals contamination, 24(6.0%) had fair knowledge while 8 (2.0%) had good knowledge.

To launch educational programs and public health initiatives, it is crucial to evaluate people's knowledge of the health issues connected with heavy metal pollution [48]. The majority of people living in the study communities are generally ignorant of the hazardous properties of heavy metals and how they affect human health. Low levels of literacy, the makeup of the communities, and a lack of exposure to information/educational programs on heavy metal toxicity may all contribute to the respondents' lack of understanding, as seen in this study. The negative health effects of heavy metal poisoning of groundwater are also largely ignored by respondents. This is consistent with the findings of a report from Cui and Forssberg [49] that the majorities of people from low socioeconomic backgrounds are not fully aware of or appropriately informed about the effects of heavy metal poisoning in groundwater.

3.6 Information on respondents' health status

The following ailments were experienced by the respondents in the last 6 months: frequent watery stool; 31(7.8%), difficulty in breathing; 14(3.5%), and skin infection; 10(2.5%). The respondents reported the following occurrences in their household in the last 1 year: still birth; 10(2.5%), stunted growth in child; 8(2.0%), and death due to cancer; 2(0.5%) in **Table 9**. According to the study's findings, a sizable percentage of respondents from the different chosen wards in Sagamu LGA, Ogun State, Nigeria, stated that they had no health issues in the previous 6 months. However, few people reported having frequent intestinal problems and having watery

Variables	Frequency	Percentage
Which of these ailments have you experienced in the last semester?		
Frequent watery stooling	31	7.8
Difficulty in breathing	14	3.5
Skin infection	10	2.5
None	345	86.2
Total	400	100.0
Have you experienced still births in your household in the last 1 year?		
Still birth	10	2.5
None	390	97.5
Total	400	100.0
Did any child in your household showed signs of stunted growth in the last 1 year?		
Yes	8	2.0
No	392	98.0
Total	400	100.0
Witnessed any death due to cancer in your household in the last 1 year?		
Yes	2	0.5
No	398	99.5
Total	400	100.0

Table 9.
Information on respondents health status.

stools, which are signs of diarrhea, typhoid fever, and other gastrointestinal-related illnesses. A negligible percentage of respondents also mentioned having children with cancer, stillbirths, and stunted growth in their households. This can be a result of heavy metal pollution from industrial effluent contaminating the groundwater. This claim is supported by the reports of numerous authors from various states in Nigeria who detailed the detrimental effects of heavy metals in drinking water from groundwater sources on human health. These reports were made by Jatau et al. [50] in Kaduna South Industrial Area; Yaya and Ahmed [51] in the Federal Capital Territory, Abuja; Nwankwoala et al. [52] in the Bayelsa town of Yenegoa; Mile et al. [53] in Makurdi and sub-urban; and Ocheri et al. [54].

4. Implications of findings and recommendations

The findings of this study have several implications for public health in terms of prevention of heavy metals contamination of groundwater by industrial activities, safeguarding the health and prolonging the lives of the local residents. Consequential to this, is the pivotal role of health promotion and education, as it aims to improve and modify people's knowledge, attitudes, and practices in order to help them achieve the highest level of salubrity through effective factual information transmission. Findings from this study has shown that there is heavy metals contamination of groundwater source in the selected communities above the permissible limits as recommended by WHO and NSDWQ, and there is reported health consequences among the populaces due to this contamination. Based on these findings, the following recommendations are suggested:

1. All industries effluents and wastewater should be properly treated to removed heavy metals and other contaminants before discharging into the environment,

2. There is need for widespread campaign and awareness program to inform the local residents on the dangers and toxicities associated with consumption of heavy metals contaminated groundwater,

3. Governments should provide facilities that would be used to treat groundwater for local consumption,

4. There is also need to educate agriculturalists and farmers on the appropriate use of pesticides and fertilizers on farmland to prevent leaching of wastewater to groundwater. Additionally, there is need for farmlands to be site away from industrial regions and areas disposed to pollution,

5. Finally, there is an urgent need for the Federal Government of Nigeria through the Ministry of water resources to embark on periodic and consistent monitoring of the underground aquifers nationwide.

5. Conclusion

This chapter present specific details regarding the condition of groundwater's quality located in communities within the industrial hub of Shagamu, Ogun State, in southwestern Nigeria with the health implications on the local residents. Since the water in the communities' hand-dug wells and boreholes has a mean pH value that is somewhat acidic, it is an appropriate medium for the breakdown of heavy metals. Additionally, heavy metals were found in the analyzed groundwater, which may be related to the soil's ability to absorb hazardous waste and untreated effluents from nearby industrial activity into subterranean aquifers. Furthermore, the levels of Pb, Cd, and Cr were higher than the permitted limits set by the WHO and NSDWQ, indicating that the people may be at risk for harmful effects from heavy metals. This proposition is further accentuated by some of the health problems such as diarrhea, stunted growth, still birth and cancer reported by the respondents. Nevertheless, the community residents in the study areas have poor knowledge on heavy metal toxicity and their inimical effects on human health.

Acknowledgements

The contribution and technical assistance of Mr. Femi Oyediran, Managing Director, Environmental Laboratories Limited and Mr. Toyin Bawala during the data collection phase and the laboratory examination of the samples are greatly valued.

Author contributions

Conceptualization, O.T.O., and A.B.T.; Methodology, O.T.O., and A.B.T.; Validation, O.T.O.; Resources, A.B.T.; Investigation, O.T.O., and A.B.T.; Data Curation, A.B. T.; Data Analysis, O.T.O., A.B.T., and S.A.A.A.; Writing—Original Draft Preparation, A.B.T., S.A.A.A. and S.A.M.A.; Writing—Review and Editing, O.T.O., S.A.A.A., S.A. M.A., and S.A.E.D.; All authors read and approved the final version of the book chapter.

Conflict of interest

The authors declare no conflict of interest.

DOI: http://dx.doi.org/10.5772/intechopen.109575

A. Appendices

A.1. Appendix I

S_N	Location	Street	Easting	North	Watertype	Lead	Lead_Conc	Cadmium	Cadmium_Co	Arsenic	Arsenic_Co	Chromium	Chromium_C
1	Eweruku	Eweruku	558,397	745,782	Well	0.13	POS	<0.01	NEG	<0.0001	NEG	<0.001	NEG
2	Eweruku	Ita Sanni	558,240	745,635	B/H	0.10	POS	0.01	POS	0.0033	NEG	<0.001	NEG
3	Kamalo	Kamalo	558,167	745,519	Well	0.14	POS	<0.01	NEG	<0.0001	NEG	<0.001	NEG
4	Kamalo	Ewu-Jagun	558,116	745,471	B/H	0.23	POS	0.11	POS	0.0010	NEG	0.036	NEG
5	Kamalo	Ewu-Jagun	558,012	745,605	B/H	0.19	POS	0.04	POS	<0.0001	NEG	0.025	NEG
6	Kamalo	Jagun	557,817	745,533	B/H	0.01	NEG	0.25	POS	0.0037	NEG	0.134	POS
7	Ogijo	Monakare	558,310	744,020	B/H	0.19	POS	0.11	POS	0.0013	NEG	<0.001	NEG
8	Ogijo	Monakare	558,385	743,977	B/H	0.23	POS	<0.01	NEG	<0.0001	NEG	<0.001	NEG
9	Ogijo	Jacob Clo	558,393	743,774	B/H	0.25	POS	0.01	POS	<0.0001	NEG	<0.001	NEG
10	Ogijo	Jacob Clo	558,425	743,629	Well	0.19	POS	<0.01	NEG	<0.0001	NEG	<0.001	NEG
11	Ogijo	Opo-Noah	558,514	743,845	B/H	0.16	POS	0.10	POS	0.0019	NEG	0.011	NEG
12	Ogijo	Manakare	558,432	743,960	B/H	0.24	POS	<0.01	NEG	<0.0001	NEG	<0.001	NEG
13	Ogijo	Ita Yakub	558,139	743,838	B/H	0.31	POS	<0.01	NEG	<0.0001	NEG	<0.001	NEG
14	Ogijo	Ita Yakub	558,082	743,689	B/H	0.52	POS	<0.01	NEG	<0.0001	NEG	<0.001	NEG
15	Ogijo	Ifesowapo	557,767	743,225	B/H	0.28	POS	<0.01	NEG	<0.0001	NEG	<0.001	NEG
16	Ogijo	Ifesowapo	558,088	743,291	B/H	0.52	POS	<0.01	NEG	<0.001	NEG	<0.001	NEG
17	Ogijo	Ifesowapo	557,933	743,225	B/H	<0.01	NEG	0.15	POS	<0.0001	NEG	0.021	NEG
18	Ogijo	Africa Re	557,641	743,013	B/H	0.61	POS	0.32	POS	0.0025	NEG	0.042	NEG
19	Ogijo	Spring Gl	557,585	743,008	B/H	0.65	POS	<0.01	NEG	<0.0001	NEG	<0.001	NEG
20	Ogijo	Seidu Str	557,337	743,095	B/H	0.30	POS	0.18	POS	0.0037	NEG	0.028	NEG

S_N	Location	Street	Easting	North	Watertype	Lead	Lead_Conc	Cadmium	Cadmium_Co	Arsenic	Arsenic_Co	Chromium	Chromium_C
21	Ogijo	Albarika	557,267	742,684	B/H	<0.01	NEG	<0.01	NEG	<0.0001	NEG	<0.001	NEG
22	Ogijo	Pz Estate	557,388	742,620	B/H	<0.01	NEG	<0.01	NEG	<0.0001	NEG	<0.001	NEG
23	Ogijo	Jumarty P	557,398	742,581	B/H	0.16	POS	<0.01	NEG	<0.0001	NEG	<0.001	NEG
24	Ogijo	Sah Boliz	557,322	742,646	B/H	0.02	POS	<0.01	NEG	<0.0001	NEG	<0.001	NEG
25	Ogijo	Goodness	557,163	742,701	B/H	0.40	POS	<0.01	NEG	<0.0001	NEG	<0.001	NEG
26	Likosi	Likosi Op	559,787	747,741	B/H	0.8900	POS	<0.01	NEG	<0.0001	NEG	<0.001	NEG
27	Likosi	Likosi Op	559,787	747,749	B/H	0.5700	POS	0.0500	POS	0.0051	NEG	0.0310	NEG
28	Likosi	NNPC Depo	559,994	747,695	B/H	0.7500	POS	0.0800	POS	0.0012	NEG	0.0770	POS
29	Sabo2 Ce	Met	567,667	753,352	B/H	1.0400	POS	0.0700	POS	0.0016	NEG	0.0130	NEG
30	Sabo2 Ce	Basid San	567,759	753,456	B/H	1.0300	POS	<0.01	NEG	0.0014	NEG	0.011	NEG
31	Sabo 2 Ce	Beside Ha	568,101	753,645	Well	1.0500	POS	<0.01	NEG	<0.0001	NEG	<0.001	NEG
32	Sabo 2 Ce	Oposite L	568,024	753,678	Well	2.2000	POS	<0.01	NEG	<0.0001	NEG	<0.001	NEG
33	Sabo 2 Ce	Inside La	568,513	753,800	B/H	0.7700	POS	<0.01	NEG	<0.0001	NEG	<0.001	NEG
34	Sabo 2 Ce	Shagamu I	568,618	753,787	B/H	3.2600	POS	<0.01	NEG	0.0089	NEG	<0.001	NEG
35	Sabo 2	Remo Divi	569,583	754,457	294 B/H	0.0300	POS	<0.01	NEG	<0.0001	NEG	<0.001	NEG
36	Sabo 2	Bureau of	569,462	754,478	295B/H	<0.01	NEG	<0.01	NEG	<0.0001	NEG	<0.001	NEG
37	Sabo 2	Orirentem	569,826	754,641	B/H	<0.01	NEG	0.0300	POS	<0.0001	NEG	0.0110	NEG
38	Kara Sabo	Oak Heigh	569,859	754,650	B/H	0.16	POS	0.15	POS	0.0070	NEG	0.062	POS
39	Sabo 1 Ka	249 Akari	569,887	754,687	B/H	0.2100	POS	0.3600	POS	0.0001	NEG	0.1470	POS
40	Sabo 1 Ka	Vkay Hote	569,949	754,699	Well	0.2900	POS	0.12	POS	<0.0001	NEG	0.1030	POS
41	Sabo 1 Ka	Amen Hosp	569,989	754,756	Well	0.0700	POS	0.2700	POS	0.0021	NEG	0.0720	POS
42	Sabo 1 Ka	Kara	570,079	754,818	B/H	0.1700	POS	0.2000	POS	0.0032	NEG	0.0350	NEG
43	Sabo 1 Ka	5 Musubau	570,133	754,708	Well	0.3500	POS	0.4300	POS	0.0041	NEG	0.1390	POS
44	Sabo 1 Ka	10 Musuba	570,164	754,673	29 Well	0.9700	POS	0.1600	POS	0.0017	NEG	0.0530	POS

S_N	Location	Street	Easting	North	Watertype	Lead	Lead_Conc	Cadmium	Cadmium_Co	Arsenic	Arsenic_Co	Chromium	Chromium_C
45	Sabo 1 Ka	Orepitan	570,063	754,625	301 Well	0.53	POS	0.26	POS	0.0011	NEG	0.124	POS
46	Sabo 1 Ka	Kara	570,148	754,921	302 B/H	1.1200	POS	0.1200	POS	<0.0001	NEG	0.0910	POS
47	Sabo 1 Ka	5 Kolawol	570,138	755,066	303 B/H	1.5500	POS	0.22	POS	0.0013	NEG	0.215	POS
48	Sabo 1 Ka	10 Kolawo	570,052	755,066	304 Well	3.12	POS	<0.01	NEG	0.0061	NEG	0.029	NEG
49	Sabo 1 Ka	243 Akari	570,236	754,900	305 Well	0.29	POS	<0.01	NEG	<0.0001	NEG	<0.001	NEG
50	Sabo 1 Ka	237Akarig	570,276	754,953	Well	0.18	POS	<0.01	NEG	<0.0001	NEG	<0.001	NEG

A.2. Appendix II: Results of pH, conductivities, temperature and TDS

S/N	Water type	pH	Conductivity (μ^2/cm)	TDS (mg/L)	Temp. (°C)
001	Well	4.5	0090	0040	27.7
002	B/H	5.5	0030	0010	30.5
003	Well	4.1	0060	0020	30.4
004	B/H	4.5	0030	0010	31.4
005	B/H	5.0	0040	0010	29.0
006	B/H	4.7	0090	0030	29.3
007	B/H	4.4	0030	0010	28.4
008	B/H	4.5	0020	NIL	29.4
009	B/H	4.5	0030	0010	28.7
010	Well	4.5	0020	NIL	31.0
011	B/H	4.4	0020	NIL	30.8
012	B/H	4.2	0020	NIL	31.6

S/N	Water type	pH	Conductivity (μ^2/cm)	TDS (mg/L)	Temp. (°C)
013	B/H	4.0	0020	NIL	28.2
014	B/H	4.6	0020	NIL	28.5
015	B/H	4.4	0020	NIL	29.3
016	B/H	4.5	0030	0010	34.5
017	B/H	4.3	0020	NIL	30.7
018	B/H	4.0	0020	NIL	32.0
019	B/H	4.1	0020	NIL	31.1
020	B/H	5.1	0030	0010	32.2
021	B/H	4.4	0030	0010	28.5
022	B/H	4.7	0010	NIL	31.8
023	B/H	4.1	0010	NIL	30.5
024	B/H	3.8	0020	NIL	30.5
025	B/H	4.8	0030	0010	28.9
026	B/H	4.8	0030	0010	27.1
027	B/H	6.0	0440	0210	30.0
028	B/H	7.0	0520	0250	32.2
029	B/H	7.5	0530	0130	31.0
030	B/H	7.0	0280	0070	31.0
031	B/H	5.7	0160	0050	29.4
032	B/H	5.9	0470	0230	29.2
033	B/H	7.6	0480	0260	31.1
034	B/H	7.9	0050	0200	30.1
035	B/H	7.7	0100	0240	29.2

DOI: http://dx.doi.org/10.5772/intechopen.109575

S/N	Water type	pH	Conductivity (μ^2/cm)	TDS (mg/L)	Temp. (°C)
036	B/H	6.0	0180	0230	28.8
037	B/H	4.9	0240	0020	29.6
038	B/H	4.9	0660	0040	32.8
039	B/H	4.7	0430	0080	29.0
040	Well	5.3	0210	0110	29.8
041	B/H	5.7	0030	0320	32.1
042	B/H	5.8	0040	0210	29.4
043	Well	5.8	0710	0100	30.1
044	Well	5.1	0750	0360	32.7
045	B/H	6.0	0450	0220	29.0
046	B/H	5.6	0360	0170	32.1
047	B/H	4.3	0290	0140	31.2
048	Well	5.0	0620	0300	31.9
049	Well	4.2	0400	0190	32.9
050	Well	4.0	0420	0200	32.1

A.3. Appendix III: GIS map of Ogijo/Likosi and Sabo 1 and 2

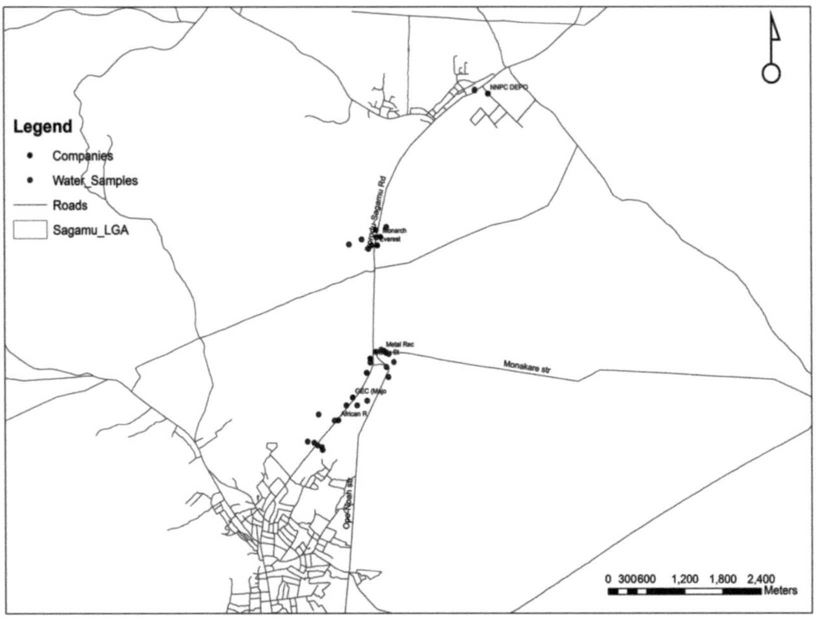

Figure A1.
Map showing sample locations in Ogijo/Likosi ward 10.

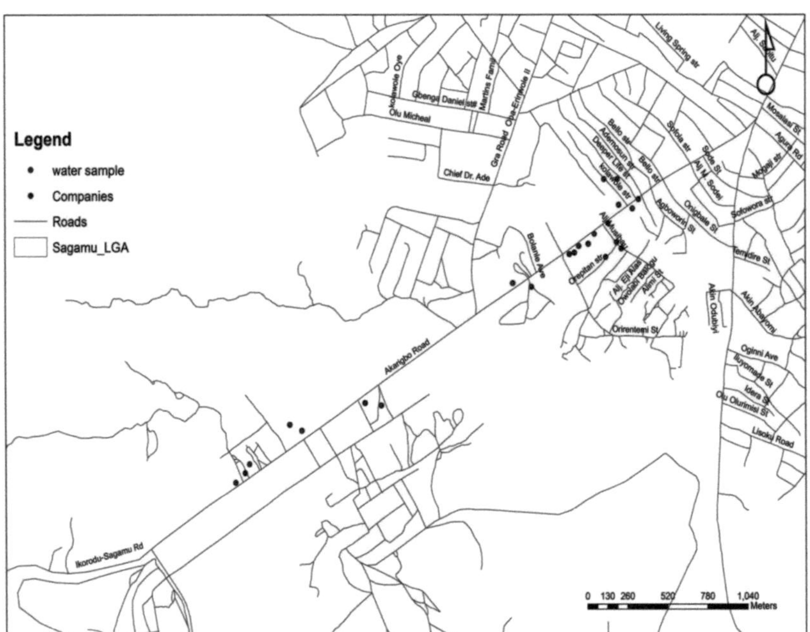

Figure A2.
Map showing sample locations in Sabo ward 4 and 5.

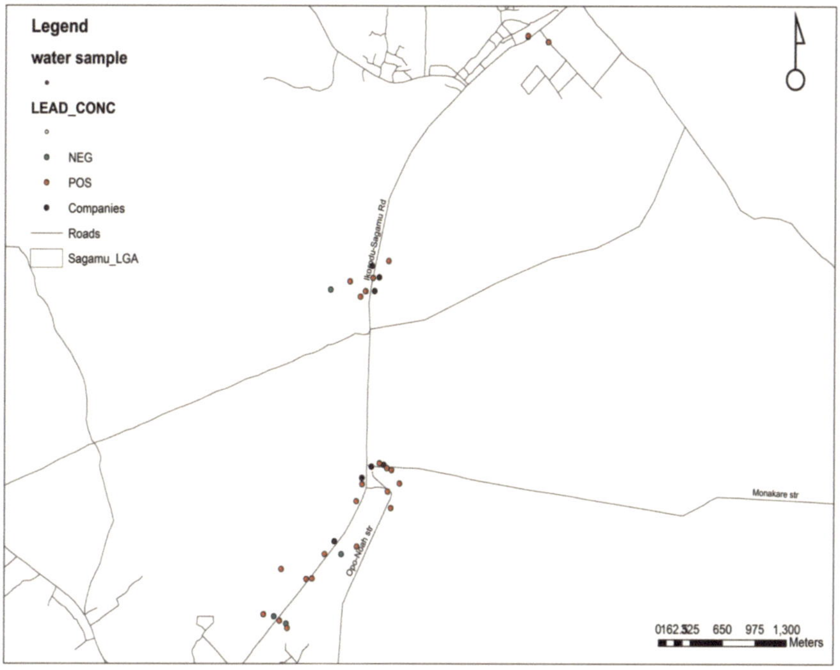

Figure A3.
Map showing lead concentration distribution in Ogijo/Likosi ward 10.

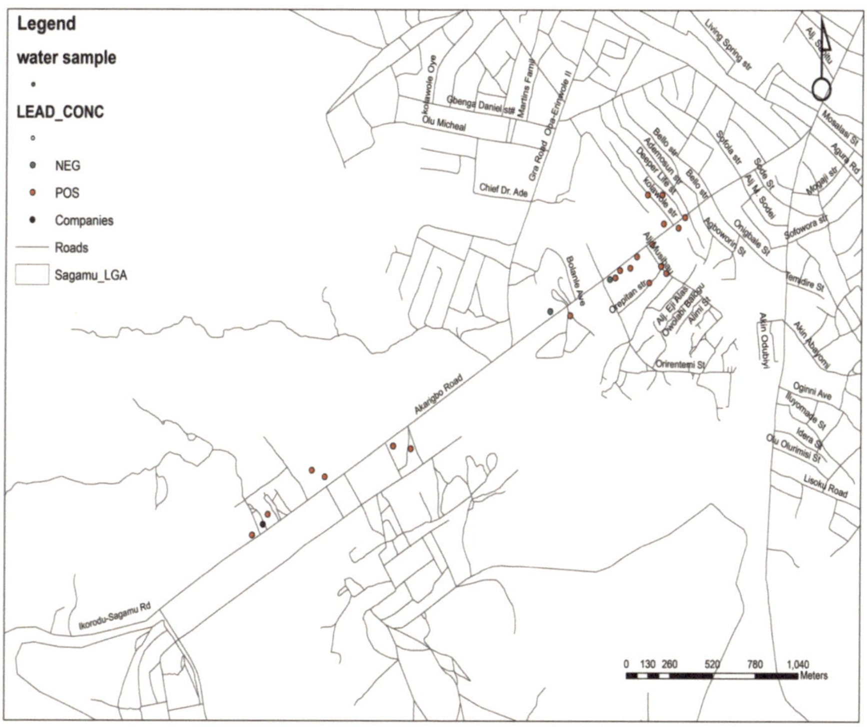

Figure A4.
Map showing lead concentration distribution in Sabo ward 4 and 5.

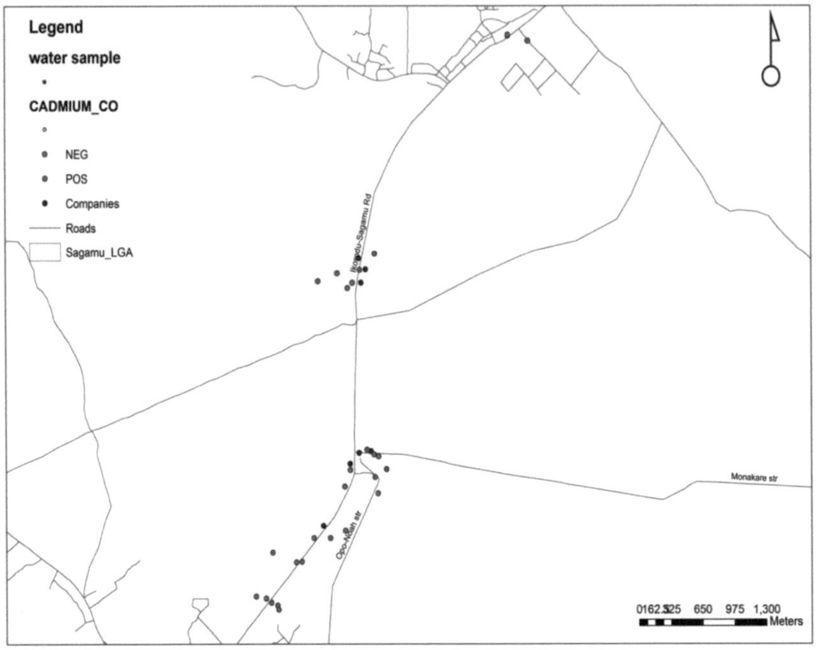

Figure A5.
Map showing cadmium concentration distribution in Ogijo/Likosi ward 10.

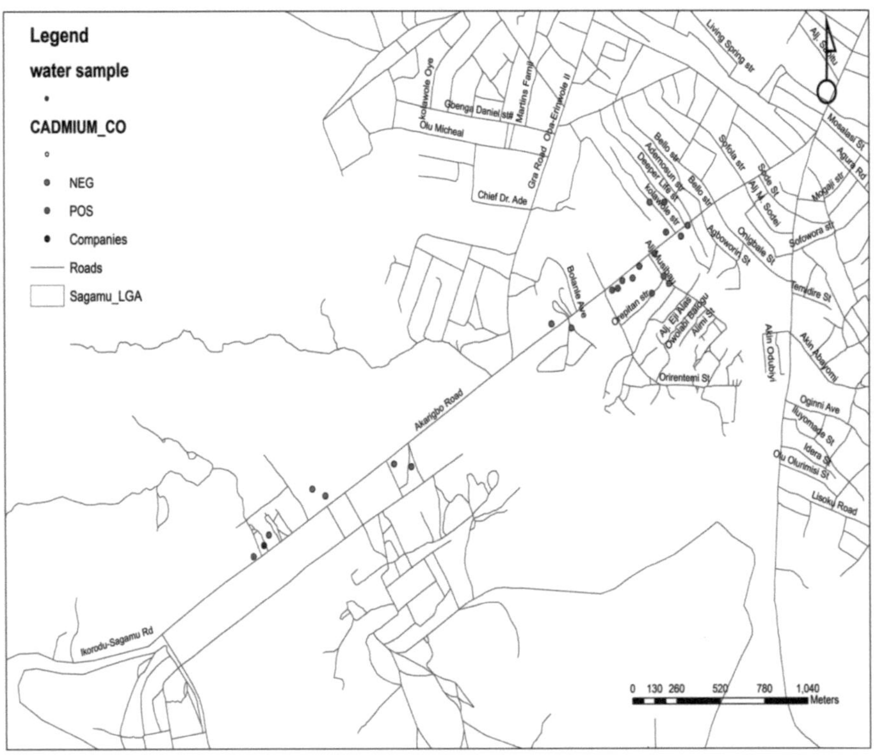

Figure A6.
Map showing cadmium concentration distribution in Sabo ward 4 and 5.

DOI: http://dx.doi.org/10.5772/intechopen.109575

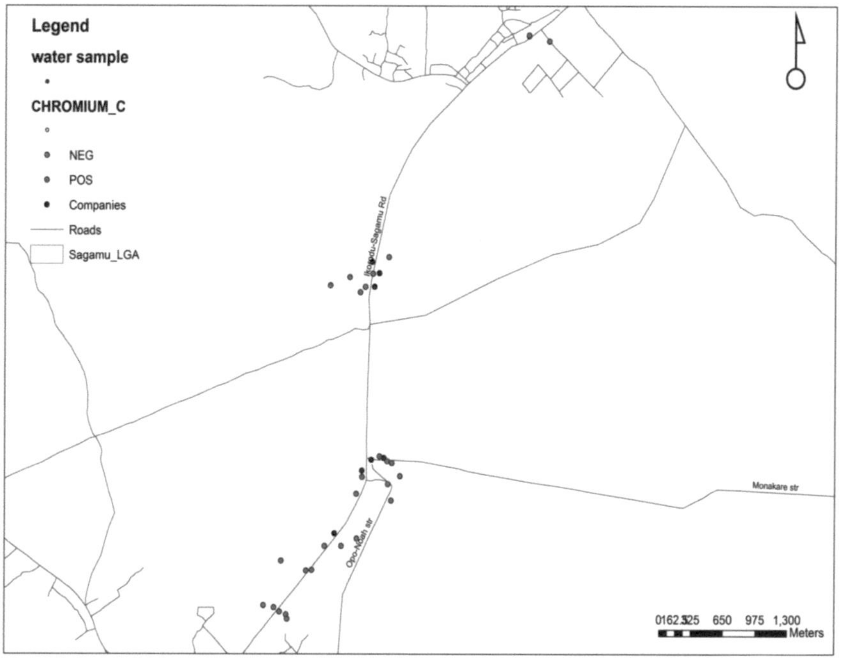

Figure A7.
Map showing chromium concentration distribution in Ogijo/Likosi ward 10.

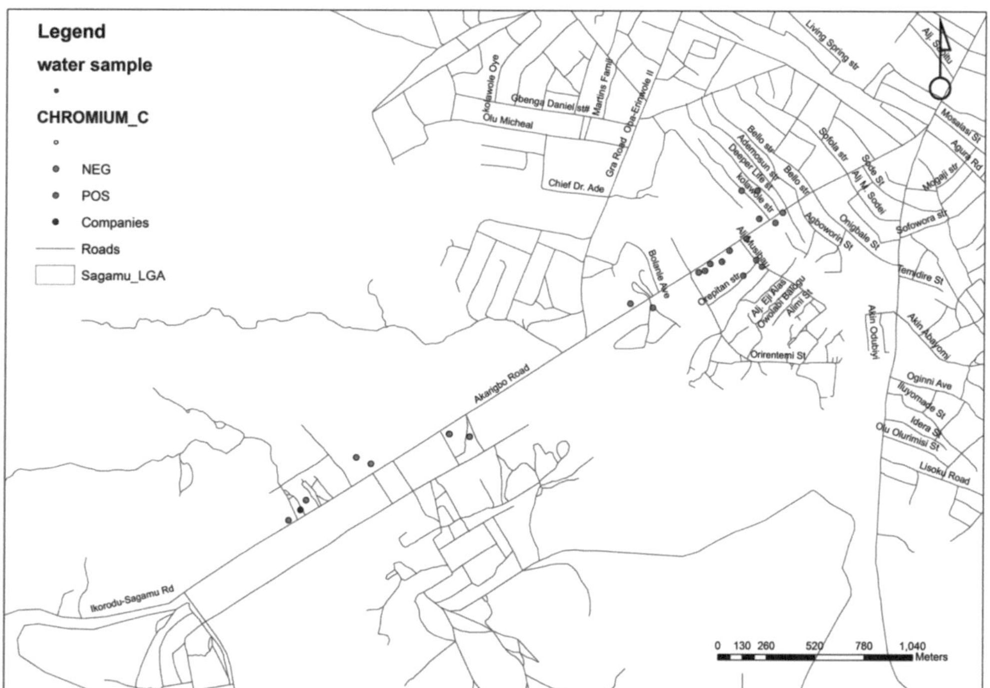

Figure A8.
Map showing chromium concentration distribution in Sabo ward 4 and 5.

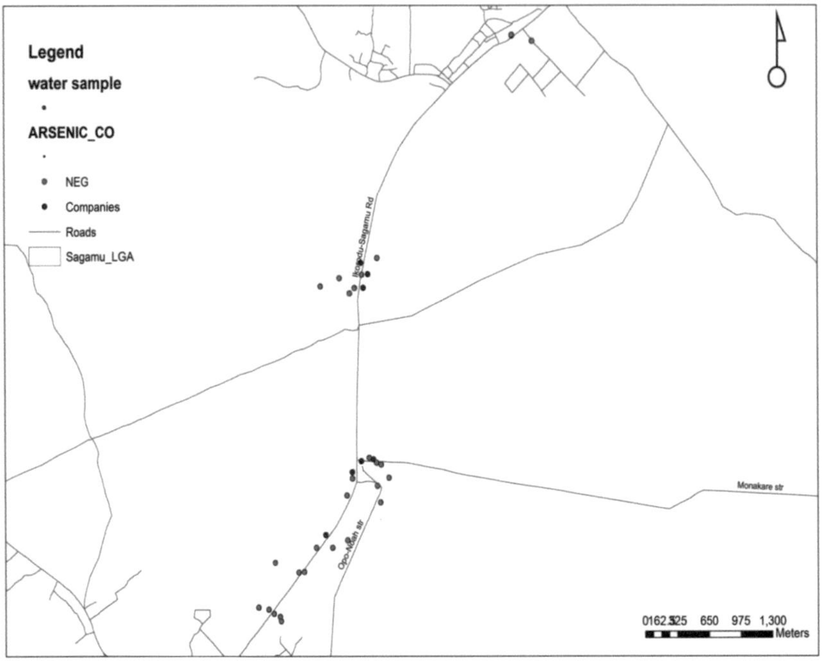

Figure A9.
Map showing arsenic concentration distribution in Ogijo Likosi ward 10.

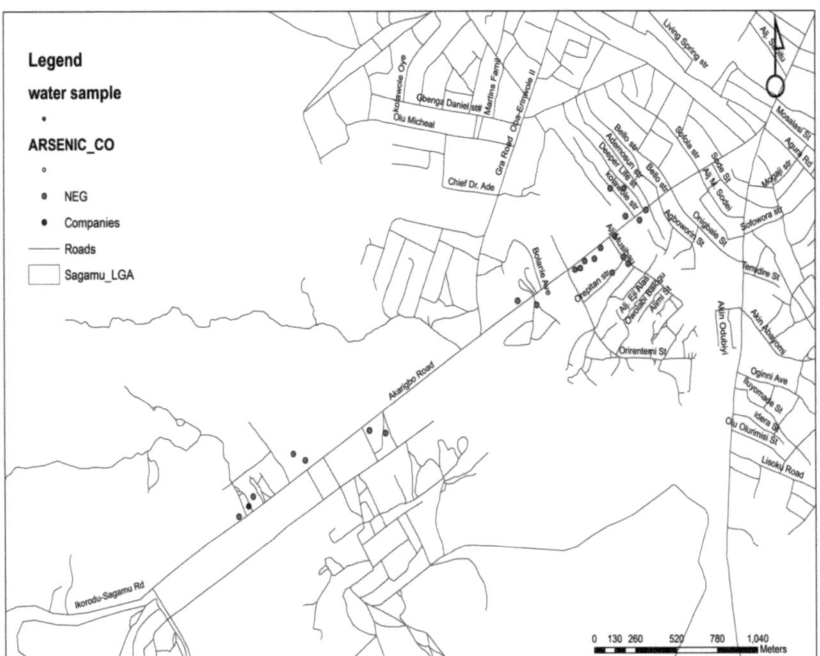

Figure A10.
Map showing arsenic concentration distribution in Sabo ward 4 and 5.

Author details

Titus Oladapo Okareh[1], Adewale Allen Sokan-Adeaga[1,2]*, Tosin Akin-Brandom[3], Micheal Ayodeji Sokan-Adeaga[4] and Eniola Deborah Sokan-Adeaga[5]

1 Faculty of Public Health, Department of Environmental Health Sciences, College of Medicine, University of Ibadan, Ibadan, Oyo State, Nigeria

2 Faculty of Public Health, Department of Environmental Health Sciences, College of Medicine, Lead City University, Ibadan, Oyo State, Nigeria

3 Department of Public Health, School of Public and Allied Health, Babcock University, Ilishan Remo, Ogun State, Nigeria

4 Faculty of Clinical Sciences, Department of Community Health and Primary Health Care, College of Medicine, University of Lagos, Lagos, Nigeria

5 Faculty of Basic Medical Sciences, Department of Physiology, College of Health Sciences, Ladoke Akintola University of Technology (LAUTECH), Ogbomosho, Oyo State, Nigeria

*Address all correspondence to: sokanadeaga.adewaleallen@yahoo.com

References

[1] Yadav KK, Gupta N, Kumar V, Arya S, Singh D. Physico-chemical analysis of selected ground water samples of Agra city, India. Recent Research in Science and Technology. 2012;**4**(11):51-54. Available from: https://updatepublishing.com/journal/index.php/rrst/article/view/989

[2] Pandey SK, Tiwari S. Physicochemical analysis of groundwater of selected area of Ghazipur city—A case study. Natural Science. 2009;7(1):17-20. ISSN: 1545-074

[3] Ullah R, Naseem R, Qadir A. Assessment of groundwater contamination in an industrial city, Siakolt, Pakistan. African Journal of Environmental Science and Technology. 2009;**3**(12):429-446

[4] Al-Bahry SN, Elshafie AE, Victor R, Mahmoud IY, Al-Hinai JA. Opportunistic pathogens relative to physicochemical factors in water storage tanks. Journal of Water and Health. 2011;**9**:382-393

[5] Sokan-Adeaga MA, Sokan-Adeaga AA, Sokan-Adeaga ED. Characterization and antibiotics sensitivity pattern of Enterobacteriaceae in Obafemi Awolowo University (OAU) sewage oxidation pond. Global Journal of Medical Research (C): Microbiology and Pathology (GJMRA). 2018;**18**(1):7-17. DOI: 10.17406/GJMRACVOL18IS1PG7. Online ISSN: 2249-4618 and Print ISSN: 0975-5888

[6] Olanrewaju JA, Akinte DO, Sokan-Adeaga AA, Sokan-Adeaga MA. A survey on faecal management practices and associated health impacts among residents in selected sub-urban communities in Ibadan, Nigeria. Journal of Applied Sciences. 2022;**22**(3):107-116. DOI: 10.3923/jas.2022.107.116. ISSN: 1812-5654

[7] Nester EW, Anderson DG, Roberts CE, Pearsall NN, Nester MT. Microbiology: A Human Perspective. 3rd ed. New York; United States: McGraw-Hill Companies Inc; 2001. pp. 787-798

[8] Felfoldi T, Heeger Z, Vargha M, Marialigeti K. Detection of potentially pathogenic bacteria in the drinking water distribution system of a hospital in Hungary. Clinical Microbiology and Infection. 2010;**16**:89-92

[9] Marcovecchio JE, Botte SE, Freije RH. Heavy metals, major metals, trace elements. In: Nollet LM, editor. Handbook of Water Analysis. 2nd ed. London: CRC Press; 2007

[10] Underwood EJ. Trace Elements in Humans and Animals' Nutrition. 3rd ed. New York: Academic Press; 1956

[11] Belkhiri L, Tiri A, Mouni L. Assessment of heavy metals contamination in groundwater: A case study of the south of Setif area, East Algeria. In: Komatina D, editor. Achievements and Challenges of Integrated River Basin Management. London, UK: IntechOpen; 2018. DOI: 10.5772/intechopen.75734

[12] Jaishankar M, Mathew BB, Shah MS, et al. Biosorption of few heavy metal ions using agricultural wastes. Journal of Environment Pollution and Human Health. 2014;**2**(1):1-6

[13] Sokan-Adeaga AA, Sokan-Adeaga MA, Sokan-Adeaga ED, Okareh TO, Edris H. Chemobiokinetics, biotoxicity and therapeutic overview of selected heavy metals poisoning: A review. Biodiversity International Journal. 2020; **4**(5):211-222. DOI: 10.15406/bij.2020.04.00187. eISSN: 2575-906X

[14] Ojekunle ZO, Adeyemi AA, Taiwo AM, Ganiyu SA, Balogun MA. Assessment of physicochemical characteristics of groundwater within selected industrial areas in Ogun State, Nigeria. Environmental Pollutants and Bioavailability. 2020;**32**(1):100-113. DOI: 10.1080/26395940.2020.1780157

[15] Scott A. Environmental impact of small scale industries in third world. In: UK Global Environmental Change Programme, ESRC, Knowledge Base; 1998

[16] https://ogunstatebiz.tripod.com/adoodootalocalgovernment/id4.html

[17] World Health Organisation (WHO). Guidelines for Drinking-Water Quality. 4th ed. Geneva: World Health Organization; 2011. p. 340. ISBN: 9789241548151. Available from: http://www.who.int [Cited: 24 November 2022]

[18] Standard Organisation of Nigeria (SON). Nigerian Standard for Drinking Water Quality (NSDWQ). NIS 554. Lagos: Standard Organization of Nigeria; 2008. p 30

[19] Okareh OT, Akin-Brandom T, Sokan-Adeaga AA. Spatial distribution of heavy metals contamination of groundwater in neighborhood communities of Shagamu Industrial Layout, Nigeria. Global Journal of Science Frontier Research (H): Environment and Earth Science (GJSRF). 2018;**18**(2):21-38. DOI: 10.17406/GJSFRVOL18IS1PG21. Online ISSN: 2249-4626 and Print ISSN: 0975-5896

[20] Asamoah DN, Amorin R. Assessment of the quality of bottled/sachet water in the Tarkwa-Nsuaem Municipality (TM) of Ghana. Journal of Applied Science, Engineering and Technology. 2011;**3**(5):377-385

[21] Meinhardt PL. Recognizing waterborne disease and the health effects of water contamination: A review of the challenges facing the medical community in the United States. Journal of Water and Health. 2006;**4**(S1):27-34

[22] Longe EO, Balogun MR. Groundwater quality assessment near a municipal landfill, Lagos, Nigeria. Research Journal of Applied Sciences, Engineering and Technology. 2010;**2**(1):39-44

[23] Rahim A, Hussain S. Assessment of groundwater quality of Beed City, Maharashtra, India. Oriental Journal of Chemistry. 2011;**27**(3):1273-1275

[24] Ganiyu SA, Badmus BS, Olurin OT, Ojekunle ZO. Evaluation of seasonal variation of water quality using multivariate statistical analysis and irrigation parameter indices in Ajakanga area, Ibadan, Nigeria. Applied Water Science. 2018;**8**(1):35. DOI: 10.1007/s13201-018-0677-y

[25] Subramani T, Elango L, Damodarasamy SR. Groundwater quality and its suitability for drinking and agricultural use in Chithar River Basin, Tamil Nadu, India. Environmental Geology. 2005;**47**(8):1099-1110

[26] Adebayo AS, Ariyibi EA, Awoyemi MO, Onyedim GC. Delineation of contamination plumes at Olubonku Dumpsite using geophysical and geochemical approach at Ede Town, Southwestern Nigeria. Geosciences. 2015;**5**(1):39-45. DOI: 10.5923/j.geo.20150501.05

[27] Adeolu AO, Oriaku AV, Adewumi GA, Otitoloju AA. Assessment

of groundwater contamination by leachate near a municipal solid waste landfill. African Journal of Environmental Science and Technology. 2011;**5**(11):933-940

[28] Wang G, Su MY, Chen YH, Lin FF, Luo D, Gao SF. Transfer characteristics of cadmium and lead from soil to the edible parts of six vegetable species in southeastern China. Environmental Pollution. 2006;**144**:127-135

[29] DeZuane J. Handbook of Drinking Water Quality. Hoboken: Wiley; 1997

[30] Popoola LT, Yusuff AS, Aderibigbe TA. Assessment of natural groundwater physico-chemical properties in major industrial and residential locations of Lagos metropolis. Applied Water Science. 2019;**9**:191. DOI: 10.1007/s13201-019-1073-y

[31] Mugica V, Maubert M, Terresa M, Munoz J, Rico E. Temporal and spatial variation of metal content in TSP and PM 10 in Mexico City during 1996–1998. Journal of Aerosol Science. 2002;**33**: 91-102

[32] SPEGH. Environmental geochemistry and health special issue. In: 18th European Conference. 2001 (Vol. 23, No. 3)

[33] Gregoriadou A, Delidou K, Dermosonoglou D, Tsoum PP, Edipidi C, Katsougiannopoulos B. Heavy metals in drinking water in Thessaloniki area, Greece. In: Proceedings of the 7th International Conference on Environmental Hazards Mitigation, Cairo University, Egypt. 2001. pp. 542-556

[34] Imam TS. Assessment of heavy metal concentrations in the surface water of Bompai-Jakara Drainage Basin, Kano State, Northern Nigeria. Bayero Journal of Pure and Applied Sciences. 2012;**5**(1):103-108

[35] Cleverland LM, Minter ML, Cobb KA, et al. Lead hazard for pregnant women and children. The American Journal of Nursing. 2008;**108**(10):40-49

[36] Chouhan S, Flora SJS. Arsenic and fluoride. Two major groundwater pollutants, India. Journal of Experimental Biology. 2010;**48**:666-678

[37] Agency for Toxic Substances and Disease Registry, ATSDR. Toxicological Profile for Arsenic. 2007

[38] World Health Organisation. Arsenic Compounds. Environmental Health Criteria 224. 2nd ed. Geneva: WHO; 2001

[39] Oladapo OT, Tosin A-B, Allen S-AA, Ayodeji S-AM. Knowledge and perceived health risks associated with heavy metals contamination in groundwater—A case study of Sagamu Local Government Area, Ogun State, Nigeria. International Journal of Health, Safety and Environment. 2020;**6**(3):506-518. ISSN: 2360-9311

[40] Dos Santos S, Adams EA, Neville G, Wada Y, de Sherbinin A, Mullin Bernhardt E, et al. Urban growth and water access in Sub-Saharan Africa: Progress, challenges, and emerging research directions. Science of the Total Environment. 2017;**607–608**:497-508

[41] Lawhon M, Nilsson D, Silver J, Ernstson H, Lwasa S. Thinking through heterogeneous infrastructure configurations. Urban Studies. 2018;**55**: 720-732

[42] Lapworth DJ, Nkhuwa DCW, Okotto-Okotto J, Pedley S, Stuart ME, Tijani MN, et al. Urban groundwater

quality in Sub-Saharan Africa: Current status and implications for water security and public health. Hydrogeology Journal. 2017;**25**: 1093-1116

[43] Foster S, Lawrence AR, Morris BL. Groundwater in Urban Development: Assessing Management Needs and Formulating Policy Strategies. Vol. 390. Washington, DC, USA: World Bank; 1998

[44] Foster S, Gathu J, Eichholz M, Hirata R. Climate change: The utility groundwater role in supply security. The Source. 2020;**18**:50-54

[45] Taylor RG, Favreau G, Scanlon BR, Villholth KG. Topical collection: Determining groundwater sustainability from long-term piezometry in Sub-Saharan Africa. Hydrogeology Journal. 2019;**27**:443-446

[46] Shamsudduha M, Taylor RG. Groundwater storage dynamics in the world's large aquifer systems from GRACE: Uncertainty and role of extreme precipitation. Earth System Dynamics. 2020;**11**:755-774

[47] Abolanle-Azeez AK, Joshua OB, Esther OA, Omos TK. Occurrence and health implications of high concentration of cadmium and arsenic in drinking water sources in selected towns of Ogun State, Southwest Nigeria. Journal of Toxicology and Environmental Health Sciences. 2010;**3**(13):385-391

[48] Ward TB, Sifonis CM. Task demands and generative thinking: What changes and what remains the same?. Journal of Creative Behavior. 1997;**31**:245-259

[49] Cui J, Forssberg E. Mechanical recycling of waste electric and electronic equipment: A review. Journal of Hazardous Materials. 2003;**99**(3): 243-263

[50] Jatau BS, Bajah I, Innocent S. Trace metal in surface and sub-surface water of Kaduna south industrial area of North Central Nigeria. Research Journal of Applied Sciences. 2008;**3**(1):81-87

[51] Yaya OO, Ahmed SD. Investigation of heavy metals in groundwater and surface water in the Federal Capital City of Abuja. In: Proceedings Annual Conference of the Nigerian Association of Hydrogeologists on Water Resources Development and Climate Change. 2010. p. 8

[52] Nwankwoala HO, Udom GT, Ugwu SA. Some heavy metal investigation in groundwater resources in Yenegoa, Bayelsa State, Nigeria. Journal of Applied Technology in Environmental Sanitation. 2011;**1**(2): 163-170

[53] Mile II, Amonum JI, Sambe NN. Heavy metals in groundwater in urban and sub-urban Makurdi, Nigeria. Pakistan Journal of Chemistry. 2013; **3**(2):87-90

[54] Ocheri MI, Odoma LA, Umar ND. Groundwater quality in Nigerian urban areas: A review. Global Journal of Science Frontier Research (H). Environmental Earth Sciences. 2014; **14**(3):1-13

Chapter 3

Cost-Effective, Sanitary Shallow Water Wells for Agriculture and Small Communities Using Mechanized Tube Well Installation

Grace L. Baldwin Kan-Uge, Tyler J. McPheron, Zackariah J. Horn and Robert M. Stwalley III

Abstract

Multiple studies have adequately demonstrated the connection between sanitary water supply for developing communities and sustainable economic growth. Unfortunately, the cost of traditional drilled water wells prevents their more rapid installation across much of the developing world. Numerous communities and agricultural areas could benefit greatly from access to groundwater less than 10 meters deep. Researchers have developed a means to mechanize shallow tube well installation to provide sanitary water wells of modest capacity. A hydraulic ram for agricultural fence post driving has been attached to a small PUP utility vehicle and repurposed to drive small diameter well pipe. This chapter will outline the water access problem from a global perspective, describe the traditional means of construction for sanitary water wells in remote areas and their relative costs, and detail the recent advancements and potential cost savings provided by a simple mechanized means to install tube wells in shallow water table areas.

Keywords: irrigation, potable water, sanitary water, shallow wells, tube well installation

1. Introduction

Globally, 1.8 billion people (22.5%) use an unimproved source of drinking water with no protection against contamination from feces. Safe drinking water, combined with good hygiene and improved general sanitation, is generally known as WASH. Improved WASH conditions could potentially prevent around 842,000 deaths each year [1]. The WASH acronym specifically stands for: safe Water Access for drinking and household use that is free from chemical and biological pollutants, Sanitation including access to a toilet (latrine) that safely separates human excreta from the environmental, and Hygiene focusing on public health and prevention of the transmission of fecal-oral diseases [2]. This chapter will examine the state of the water component of WASH programing in the developing world. Traditional techniques to access groundwater will be reviewed, and recent work using mechanized tube well installation will introduce the Well-Driver PUP technology [3]. Implementation of this technology could provide meaningful progress toward addressing the sixth U.N. Sustainable Development Goal (SDG): "To ensure availability and sustainable management of water and sanitation for all", which incidentally will help drive progress across many other SDGs [1]. People must have equitable and affordable access to safe and sufficient water, that is palatable and in sufficient quantity for both drinking and domestic purposes [4]. On-going research proposes to increase access to subsurface water by improving the operational capabilities of the Well-Driver PUP (Purdue Utility Project) vehicle [5]. Implementation and dissemination of this novel vehicle technology could improve access to safe water for drinking and domestic purposes in developing countries and can play a key role in WASH programing.

1.1 Water quality access

In developing countries, access to safe water is critical to the quality of life and the potential for economic growth. Water-related diseases pose a major risk to individuals in developing countries, through the consumption and use of unsafe and poor quality water sources [4]. Often, water sources are prone to contamination, due to the movement of contaminates through surface transport processes. Water contamination due to surface runoff, leaching, and pollution from agro-chemicals into groundwater sources can lead to increased risks of humans contracting waterborne pathogens from drinking water. Poor waste management and the inappropriate disposal of human and animal excreta can result in higher levels of contamination in water resources. The presence of excreta in water used for human consumption, often leads to serious, but preventable, diseases, such as typhoid and cholera. Water that is high in fecal coliform bacteria, which is generally greater than 99% *Escherichia coli*, indicates a level of human and animal waste contamination in the water and the possible presence of other harmful pathogens [4].

Excess fertilizer use can lead to the leaching of dissolved nitrogen through the soil profile, resulting in additions of nitrate into groundwater resources. The consumption of drinking water containing nitrate higher than 2 mg/L for adults has also been shown to lead to adverse health effects, specifically higher risks of cancers [6]. For mammals, the adverse health pathway is nitrate within drinking water increasing the production of N-nitroso compounds, which are highly carcinogenic [7]. Infants ingesting drinking water containing a high nitrate content can have low oxygen levels in their blood, leading to a potentially fatal condition, known as "blue baby syndrome." Access to water that is safe and considered of good quality is essential to overall community health and healthy living conditions for people and domestic livestock around the world.

1.2 Water quantity access

In many locations around the world, people use unsafe water sources or lack sufficient access to water for both drinking and domestic purposes, creating very unhealthy circumstances for these individuals. This is because in developing countries, clean water access is not always possible. Water resource use is often constrained due to the terrain and hydrology of a specific location. Particularly in sub-Saharan Africa, women and children often walk great distances to obtain access to water for household use. In many regions, water is carried on top of one's head, while simultaneously leading and watering

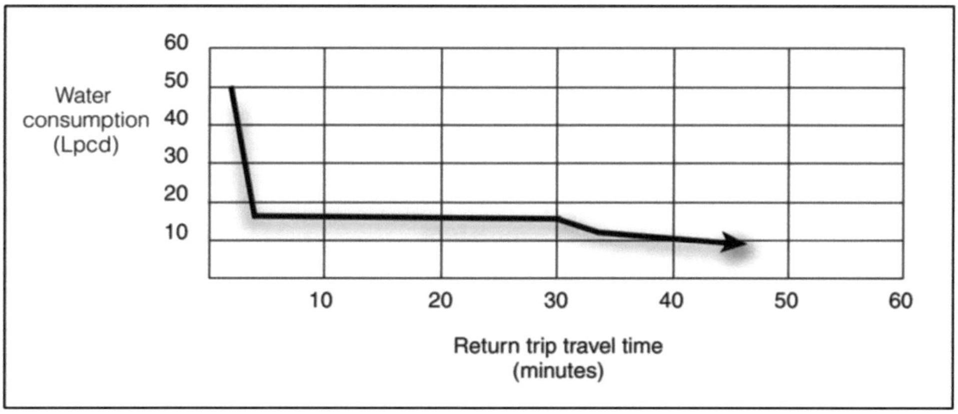

Figure 1.
The relationship between water collected, journey time, & domestic consumption [8].

livestock. According to the SPHERE humanitarian standards, for any water-based source, the distance from the household to the nearest waterpoint should not exceed 500 m, and the queue time for water sources should be no greater than 30 min [4]. In situations requiring longer travel times, individuals are far less likely to collect larger amounts of water, as seen in **Figure 1**.

These data indicate that families, especially those living farther away from a water source, will only collect the basic minimum amounts of water required for survival. Within a just society, all people would have an equitable and an affordable means of access to a sufficient supply of water that could be used for drinking, hygiene, and domestic purposes [4]. **Table 1** displays the SPHERE recommended minimum total water need for basic survival. The average water used for drinking, cooking, and personal hygiene in any household is 15 L per person per day, or for an average month of 30 days, 450 L. For a family of five, 2250 L or 2.25 ton of water would be required to meet the minimum demand for all domestic uses. Obviously, the amount of water required for an individual can vary, based-on the community and context, but human living needs require a minimum level of water of survival [4]. When this water is not located in the home and must be collected elsewhere, productive time for alternative activity is lost [9]. Women and children are disproportionately impacted by this cruel labor requirement. For women, it shortens the available time for them to be with their families, provide childcare, perform household activities, and engage in entrepreneurial enterprises. Water collection by both boys and girls, can take time away from their educations, and sometimes, it can even prevent them from attending school

Survival needs: water intake (drinking and food)	**2.5–3 liters per day**	**Depends on: the climate and individual physiology**
Basic hygiene practices	2–6 liters per day	Depends on: social and cultural norms
Basic cooking needs	3–6 liters per day	Depends on: food type, social as well as cultural norms
Total basic water needs	7.5–15 liters per day	

Table 1.
Simplified table of basic survival water needs [4].

altogether. The brutality of having to collect water and transport it on a daily basis robs both women and children of their most valuable resource [9].

1.3 Current methods of sourcing groundwater

There are three primary types of wells utilized to obtain groundwater resources for both drinking and domestic purposes. These methods include hand-dug wells, drilled wells, and driven tube wells. Hand-dug wells are constructed manually and require individuals to dig until below the water table [10]. These wells must be dug during a dry season in order to ensure that the water table is at the lowest possible level. In situations where the water table has receded below the depth of the well, the bottom of the well must be dug deeper to access the water table [11]. Hand-dug wells have a circular cross-section and should be lined with stone, brick, or tile to prevent the earthen sidewalls from collapsing inward. This type of well does not have a continuous casing and grouting, making it far more prone to contamination from surrounding surface sources [12].

The most common method for obtaining groundwater is by creating a drilled well. Unfortunately, this technique is always the most expensive of the methods considered, but you get what you pay for. This type of installation produces a dependable, long-term, sanitary well, and it is considered the "gold standard" of groundwater access. Through this means, groundwater is accessible to deeper levels than by other options. Throughout the world, the general preference for WASH programming is to install a deep-drilled well to access groundwater having a lower likelihood of containments. However, in developing countries, there are often not enough reputable companies with available drilling equipment to meet the demand for installed wells, at an affordable cost, in a timely manner. In Haiti for example, there are very few drilling companies. Even when an organization or individual has sufficient funds to install a deep well, the wait time for a drilling company to come and install a new well can be over 1 year [13].

A driven tube well can be an acceptable alternative to traditional drilled wells under certain conditions [14]. Tube wells are constructed with a well point connected to galvanized steel pipe, which serves as the well casing. The well point is sharpened and driven through the soil, accessing groundwater through a fine mesh or perforation on its circumference near the point. This type of well is commonly driven by hand, with a tripod system set-up, and installation tends to be a very labor-intensive effort. Reducing the labor element in the tube well installation process could potentially make this type of well more feasible across a considerable range of developing territory [15, 16]. The Purdue Well-Driver PUP mechanizes the tube well installation process by using a hydraulic ram. This mitigates the intensive labor component generally accompanied with driven wells and dramatically reduces the installed cost of sanitary water sources, through the improved productivity of equipment and personnel involved. The remainder of this chapter describes the current status of groundwater access, the Well-Driver PUP technology, and the economic potential of the technology.

1.4 Individual access to groundwater from wells

The physical water access point is a critical element of all wells, but it is particularly vital to community wells or those with shared access. Modern well standards require that the designer do everything possible to prevent contamination on the surface from entering the well aquifer. This criterion alone discourages the investment of effort and resources into hand-dug wells, as it is far more difficult to maintain sanitary access conditions for these types of wells. For all wells in general, surface-to-aquifer contamination results from two sources: down-the-borehole backwash and beside-the-casing downward drainage. To seal the casing from the surface, all modern wells should have adequate concrete pads surrounding the casing as it rises above the surface of the ground. The pads need to be extra strong in community well situations to withstand the burden of heavy traffic near the well-head. The pads need to be properly sloped, so the drainage is carried away from the well-head and does not accumulate nearby. The concrete pads for pumps must be left to cure adequately before the pump head and water outlet assemblies are installed. Down-the-borehole contamination is best prevented by using a check valve, sometimes called a backflow preventer, in the pump assembly. The process that must be prevented is a syphon from an above-ground water storage tank back-down into the aquifer. Any contamination present in a storage tank could potentially be injected into an underground aquifer during a syphon event at a wellhead. A community water bucket dropped into a hand-dug well poses essentially the same risk of contamination. For a drilled or driven well, a hand pump or an electric pump in the casing is the recommended means to keep a water well access draw-point sanitary and safe for all patrons.

2. Review of groundwater access technologies

This section will contain a review of groundwater access options and the types of wells and drilling methods used throughout the world. Hand-dug wells and drilled wells will be explored. The components required to install a tube well are introduced, along with a discussion of previous tube well installations using the prototype Well-Driver PUP.

2.1 Groundwater accessed drinking water options

Water sources are often classified as "improved" or "unimproved." Improved sources are piped public water into homes, public standpipes, water wells or boreholes, protected (lined) dug or hand-dug wells, protected springs, bottled water, and rainwater collection [17]. Unimproved sources are unprotected wells or springs, sachet water, vendors, tanker-trucks, and surface waters [17]. International WASH efforts tend to push communities toward the installation of improved water sources. Common components within current WASH programming efforts include community involvement through the establishment of a community-led WASH committees, the construction of new water access points, the rehabilitation of pre-existing water sources, the installations of new wells or community boreholes, small town water systems, and pipe extensions [18].

In many cases, women, poor households, and marginalized groups disproportionately experience the negative impacts of inadequate WASH resources. This primarily occurs, because these groups are more than likely to have limited access to WASH services [19–22]. Marginalized groups often have less input, both at the household and at the community level, in decision-making processes and the governance of resources relating to WASH [23]. Studies show that income, education, household size, and region are all significant predictors of access to improved water and sanitation

[24, 25]. Therefore, many WASH programs and interventions utilize the methodology of empowering beneficiaries, which increases equitable access and the sustainability of water and sanitation infrastructure solutions [26–28].

2.2 Groundwater

Water that is below the water table in soil is generally called "groundwater" [29]. This is underground water that can be removed by wells. The groundwater zone acts as a natural reservoir or system filled with fresh water. "An aquifer is a saturated bed, formation, or group of formations which yields water in sufficient quantity to be used for economic purposes" [14, 29]. Water storing formations and groundwater reservoirs are synonymous for "aquifer". There are two main types of aquifers: confined and unconfined [29]. An unconfined aquifer is where water enters from the soil surface and passes through the soil profile to enter the aquifer. A confined aquifer has an impermeable geological layer that prevents surface water from directly flowing into the aquifer. The installation of a well or borehole under these conditions includes drilling through the geological layer confining the aquifer, in order to move the water from the deep aquifer, up to some higher level. In this way, water wells are accessed for groundwater across the globe, for both drinking and domestic water uses. Properly accessed groundwater is sanitary, and it is generally sustainable.

2.3 Types of wells

In many countries, particularly those in sub-Saharan Africa, individuals obtain their drinking water from community wells, which include both protected wells and boreholes. These water access points are most commonly located outside of dwellings and are in the form of a public tap or standpipe. The terms "wells" and "boreholes" tend to be used interchangeably worldwide. A borehole is the generalized term for any narrow shaft drilled into the ground. It generally contains both a pipe casing and a well screen, to prevent the entry of soil into the water flow [30]. There are three primary methods of well construction: dug, drilled, and driven wells. Water wells can be installed either through manual methods or with powered tools [11].

2.3.1 Dug or hand-dug wells

Traditionally, dug wells are excavated by hand, using simple tools such as a pick and shovel, with a bucket on a rope to remove cuttings [11]. **Figure 2** portrays an example of a hand-dug well installation. Although some pieces of dug well construction may be mechanized to a certain degree, the process to construct this kind of well is very manual labor intensive. A dug well is excavated below the water table during the dry seasons, until the incoming water exceeds the digger's bailing rate. These wells should be circular in cross-section and lined with stones, bricks, tile, or other material to prevent the well from collapsing inward. This type of well does not have a continuous casing and grouting, making it more prone to contamination from surrounding surface sources. Dug wells have larger diameters and expose larger areas of the aquifer to the excavation. Therefore, these wells are able to obtain water from less-permeable materials, such as very fine sand, silt, or clay [12].

Most wells of this type are shallow and not able to achieve the depths that a bored or driven well can. This type of well often goes dry during droughty seasons, because the water table drops below the well bottom. It is during this type of period that

DOI: http://dx.doi.org/10.5772/intechopen.109576

Figure 2.
Cross-section view of hand digging a water well [11].

maintenance on the well can be performed. However, working on dug wells is quite risky. Someone must be lowered into the well to work. Labor under these installations is potentially dangerous, due to the high potential for cave-ins and the lack of oxygen. Since it is difficult to dig very deep, hand-dug wells generally extend no further than 30 m in depth [11].

To access more water in situations where the water table has dropped lower than the depth of the well, the bottom of the well must be excavated deeper to reach the new aquifer level. Water is typically lifted to the surface by attaching a bucket to a rope and drawing the water up by hand or crank. Unfortunately, obtaining water in this manner can also transmit bacteria into the groundwater source. Contamination of the water source is best prevented by sealing the walls, pouring a concrete apron around the base of the well, providing a raised parapet above the face around the well, using a lid over the top of the well, and utilizing a hand or electric pump to obtain water. Obviously, these features add additional costs to the well [11].

2.3.2 Drilled wells

A well drilling machine is normally referred to as a "drill rig" or just a "rig" [11]. Drilled wells are able to penetrate consolidated material and require the installation of casing and a screen to prevent the inflow of sediment and to keep the well from collapsing inward [12]. This type of well can be pushed to more than 300 m in depth. The surface area around the casing has a segmented or concrete pad that is constructed to prevent contamination by water draining from the surrounding surface downward around the outer portion of the casing. The pad is most often constructed from neat cement or bentonite clay [12]. Pads are typically left to cure for a period of time prior to well commissioning, during which a well casing cap remains on the newly drilled well to prevent contamination. After the pad has cured, the pump cap can be removed, and a pump head can be installed. Installing a pump head too soon, prior to pad curing, can lead to breakage of the concrete pad in use. Thus, it is vital to provide a proper cure time for the concrete when installing pump equipment and subjecting the pad to heavy operational loadings. Powered well drilling methods include the percussion cable tool, jetting, mud rotary, and air rotary techniques [11]. The most common powered installation methods used today are the percussion cable method and the mud and air rotary methods, but drilled wells can also be installed by hand [12, 31, 32].

2.3.3 Bored or hand augered wells

This method of drilling a well uses a small-diameter open-bottom bucket with angled teeth to manually cut into the soil. An example of this type of installation is shown in **Figure 3**. The bucket is attached to a t-shaped handle at the top through a series of steel rods, which can be rotated manually and pushed downward. As the bucket fills, the contents are lifted-out and emptied. Additional rods are added as the hole deepens. This method is sometimes used for soil sampling, in addition to shallow well construction. The diameter of the hole produced by this method is typically less than 8 cm, and the process is very depth limited, as the drilling rate and material removal are very slow. Once below the water table, it is generally problematic to go deeper, and it is difficult to prevent the hole from collapsing inward. As well, the soil profile and composition greatly affect the depth of a well that can be installed using this method. In loose silt or sand, it is possible to go up to 10 m in depth, but in more compacted soil, it would be quite difficult to reach this depth manually drilling [11]. Therefore, most drilled well installations have been adapted to utilize machinery instead of human power.

2.3.4 Percussion (cable method)

This well drilling method utilizes repeated lifts and drops of a chisel-edged bit to break-loose and pulverize material in the bottom of the hole. A small amount of water is

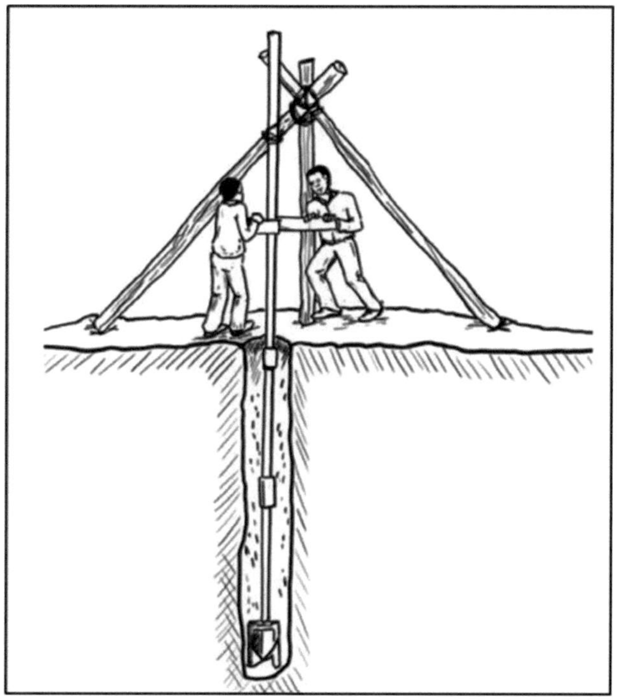

Figure 3.
Cross-section view of a hang augered water well installation [11].

added to the hole to form a slurry of the excavated material. The percussion bit is removed periodically, and a bailer lowered into the hole to remove the slurry mixture containing the excavated material. The excavated material is brought to the surface and discarded. Bailing is repeated until the hole has been thoroughly cleaned. Bailing and drilling are alternated in this fashion, until the desired depth is reached. If the hole is unstable, a casing can be lowered into the hole to prevent it from collapsing. The percussion drilling method is able to penetrate all types of materials, but in very hard stone, progress can be quite slow. The percussion technique is frequently associated with a large, truck-mounted attachments or motorized trailers, similar to that shown in **Figure 4** [31]. "The [percussion] machinery ranges from a basic skid-mounted powered winch with a tripod, to a complex set of pulleys and runs with a large mast" [11]. These larger cable tool rigs have hydraulic motors to raise and lower the mast and rotate the drums of the cable. Fewer cable tool rigs are being utilized in developed areas of the world today, because compared to hydraulic rotary drill rigs of similar size, percussion drill rigs work slower [11]. Additionally, when drilling in loose sediments, it is necessary to drive a steel pipe behind the drill bit to prevent the borehole from collapsing. The sections of this "drive casing" must be welded together going in and cut apart coming out, which requires that an arc welding and cutting torch set be available during the drilling process [11]. These additional processes are not required when using alternative well drilling technologies.

2.3.5 Jetting

This well drilling technique utilizes a high-pressure pump to force water down a drill pipe and out a small diameter nozzle, in order to make a "jet" of water that

Figure 4.
Truck-mounted cable tool rig (left) and trailer-mounted tool rig (right) [31].

loosens the soil. The return flow of water outside the drill pipe carries the cuttings up to the surface and into a settling pit. A circulation pump returns the water back-down the pipe to continue bringing more cuttings to the surface. A tripod set-up is typically used and rotated by hand to ensure a straight borehole. In addition to the water piping components, this well drilling technique requires a high-pressure water pump and two people to set-up and operate the rig. **Figure 5** illustrates this method of well drilling. Unfortunately, this method is only suitable for fine-grained and soft sediment soils, and it requires a nearby water source to supply the jet system. This well drilling technique is not suitable for gravel or in hard soil profiles [11].

2.3.6 Mud rotary

A mud rotary drilling rig includes a "jet", in combination with a larger diameter cutting bit, pre-cut and threaded lengths of steel drill pipe, a motor to turn and lift the drill pipe, and a sturdy mast to grip and support the pipe. **Figure 6** illustrates an example of this process. A mixture of bentonite clay or other materials is used in combination with water to improve the ability to lift the cuttings out of the borehole.

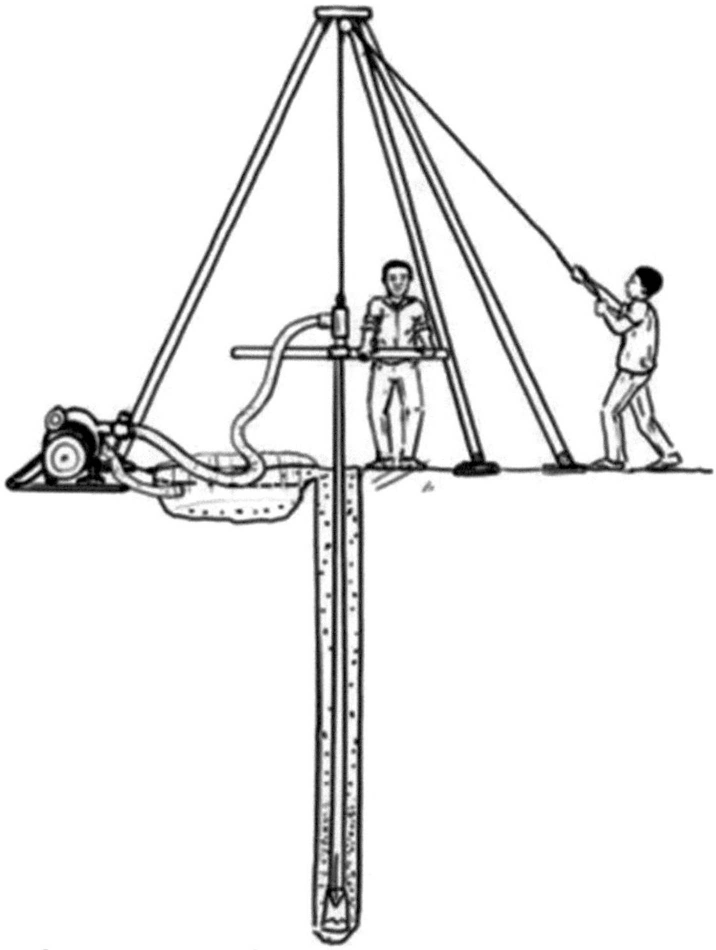

Figure 5.
Cross-section view of jetting a drilled water well [11].

This mixture is called "drilling mud" or just "mud". There are many different kinds of rotary drilling rigs, but they can be summarized into two basic set-ups: a table drive unit or a top-head drive unit. A table drive drilling rig rotates the pipe using a pipe grip and spinning mechanism near the base of the rig. A top-head drive turns the drilled pipe by way of a motor attached to the upper end of the pipe. In both set-ups, the drill pipe is also attached to a lifting mechanism that lowers and raises the pipe along the mast. A swivel on top of the pipe is present in both set-ups, allowing the drilling mud to be pumped down the drill pipe, while it is rotating.

Mud rotary well drilling is much faster than using the cable drilling technique, and mud rotary machines are capable of drilling a borehole of up to 60 cm or more in diameter. They can achieve depths of up to 60 m. In comparison with the cable drilling technique though, mud rotary rigs are more energy intensive, and they require more fuel per hour to power them. Additional components on this machine beyond a cable drilling rig include a motor to rotate the pipe column, the pipe winch, and the mud pump [11].

Figure 6.
An onsite isometric view of a mud rotary drilling rig in an operational configuration [11].

2.3.7 Air rotary

The primary difference between the air rotary drilling method and the mud rotary drilling method is that the air technique utilizes compressed air to remove the cuttings, rather than drilling mud. A type of "foam" can be added to the air stream in order to improve its effectiveness at the cuttings removal and provide additional stability to the borehole. The mechanical elements of the pipe mechanisms on the mud rotary and air rotary machines are the same. Both styles of machine can come with either a table drive unit or a top-head drive unit. Both require a pipe winch. The air rotary rig utilizes the same type of drill bits as that of a mud rig, but it also makes use of a "down-the-hole" hammer drill action. The bit used in air rotary rig operations directs a jet of compressed air to break-up rock and drill extremely fast. This type of drilling technique can be set-up very quickly, since no mud or cutting mix is utilized,

DOI: http://dx.doi.org/10.5772/intechopen.109576

Figure 7.
An air rotary drilling rig in transport configuration [31].

only compressed air. This method is able to drill much faster than other rigs of comparable size, and it creates less of a mess at the bore site. However, the air compressors utilized by air rotary rigs are generally very large, which adds additional capital cost, potential maintenance needs, and further increased fuel use [11]. An example of a typical air rotary drilling rig is shown in **Figure 7**.

2.3.8 Driven wells (tube wells)

Driven wells require the following four primary components: a well point, well point couplings, lengths of galvanized steel pipe, and a well point drive cap [10]. Each of these components is displayed in **Figures 8–11**. Galvanized pipe is required for long-term water system integrity and is commonly available [33]. Driven wells or shallow tube wells are constructed by driving a small diameter pipe into a shallow water-bearing soil profile composed of primarily sand or gravel [12, 32]. Unlike the other well construction techniques mentioned previously, material is not removed, but rather, it is forced aside during the driving process [31]. A screened well point is attached to the bottom of the casing before driving [12]. Couplings are used to connect each section of piping as needed. The drive cap is screwed onto the upper end of the section of pipe that will be driven, so as to protect the pipe threads during driving. "The drive and couplings, in addition to being heavier than standard pipe, are designed so that the pipe ends butt together inside the coupling, resulting in most of the driving force being transmitted by the ends of the pipe rather than by the threads" [31]. Driving is done by alternately raising and dropping a weight, which is used as the driving ram. A drive point and manual installation are shown in **Figure 12**. In place of only using a hammer to drive, guides can be employed to

Figure 8.
A well point for a 2″ pipe driven tube well installation [10].

Figure 9.
A drive coupling for a 2″ driven tube well stack [5].

Figure 10.
A typical unthreaded piece of potable water safe galvanized 2" steel pipe [33].

Figure 11.
A tube well installation drive cap [5].

direct the pipe during driving. This can be done either by having a guide on the outside of the pipe or the inside of the pipe as shown in **Figures 13** and **14**, respectively [31]. Prior to the Well Driver PUP, driven wells were typically installed manually or using a derrick constructed in place for the specific purpose [10].

The achievable depth to which a tube well can be driven depends on the build-up of friction between the well pipe, the material penetrated, and the transmission of the force of the driver down the length of pipe [31]. It is possible to achieve maximum depths of 25–30 m, but hard formations cannot be penetrated [31]. Tube wells are most easily installed in locations where the soil profile is mostly loose sand, and the water table is high. Locations close to a river, lake, or stream are especially good [11]. Driven wells are simple and economical to construct. Driven wells are generally not sealed at the surface using grouting material [12], and therefore, they may normally lack an adequate sanitary seal [31]. For these reasons, this type of well is more prone to contamination. However, the proper finish with an apron and riser pipe can prevent most common contamination issues [10]. Manually-driven wells are not generally able to penetrate more than 5 m below the surface [12], but this level has already been surpassed by the mechanized Well Driver PUP technology [3].

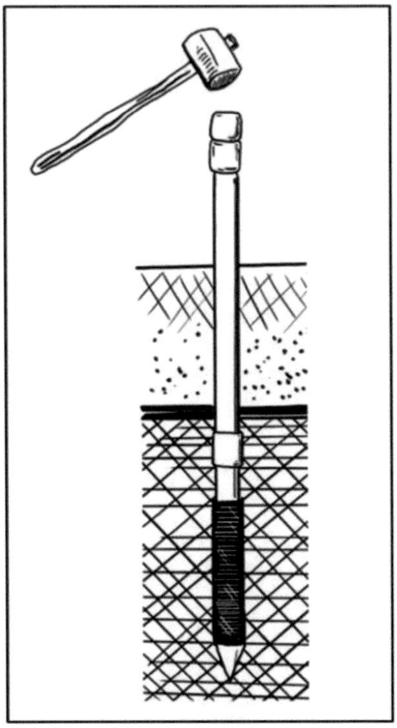

Figure 12.
A cross-section view of a manual installation of a drive tube well installation [11].

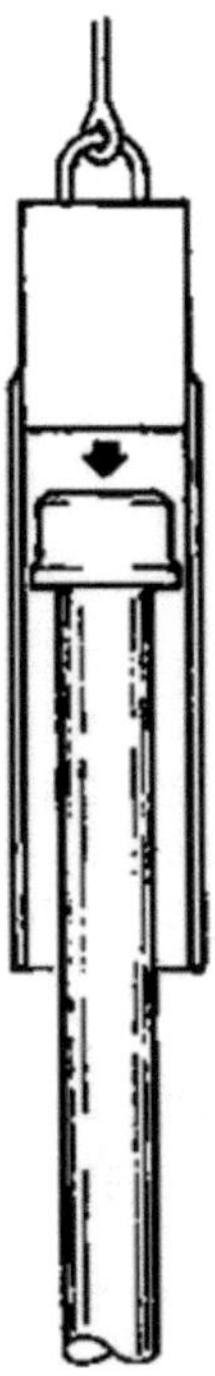

Figure 13.
Cross-section view of a device for well driving by a guiding sleeve outside of the well pipe [31].

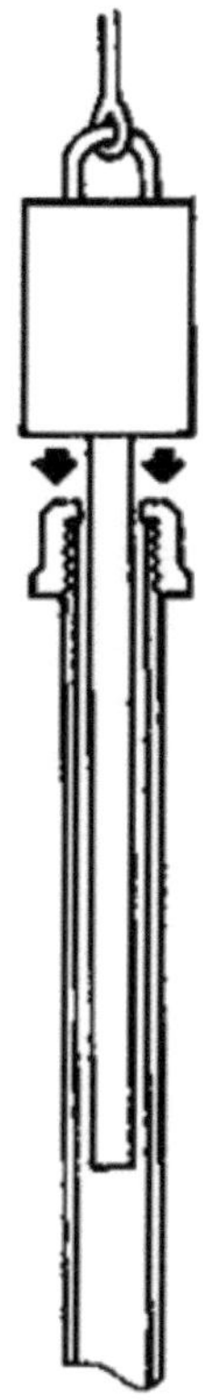

Figure 14.
Cross-section view of device for well driving by a guiding line-up probe inside of the well pipe [31].

3. Well-driver PUP

The PUP vehicle is a development of the Agricultural and Biological Engineering department at Purdue University under the direction of Dr. John Lumkes. This device is a versatile, useful, and inexpensive tool for the developing world. Researchers at Purdue have modified the basic device to install tube wells [10]. This section will introduce the PUP vehicle, the hydraulic ram attachment, the potential for impact worldwide, the components need to drive a tube well, and the results from preliminary work. A review of healthy water quality will be provided, along with an estimated cost structure for these types of installations.

3.1 Purdue utility platform (PUP) with well-driver attachment

A hydraulic post driver mated to a Purdue Utility Platform (PUP) vehicle has been designed to mechanize the process of installing driven water wells. This machine has been designated the Well-Driver PUP [10]. A PUP is a three wheeled, low-cost utility vehicle designed at Purdue University for use in developing countries [3, 34]. These vehicles are typically built in-country with minimal tooling and using only locally-sourced materials. The experience base for this vehicle is predominately in sub-Saharan African countries, mainly being Guinea, Cameroon, Nigeria, Uganda, and Kenya [34, 35]. **Figure 15** shows multiple examples of Purdue student-built PUP vehicles. These vehicles have been used previously for light commercial transportation

Figure 15.
PUP vehicles produced under the direction of Dr. John Lumkes of the Purdue University Agricultural and Biological Engineering Department [10].

purposes, including use as a small pickup truck, taxi, fire truck, or for miscellaneous hauling in areas where normal vehicles are not appropriate, due to the severe terrain and lack of road accessibility. These vehicles have even been used as ambulances to get individuals in need of medical services from extremely rural areas to hospitals and clinics. They have been used for transporting goods to and from market, as school bus alternatives, garbage collection vehicles, and light utility tractors. Tillage attachments, seeders, harvest heads, water pumps, generators, threshers, and maize grinders powered by the PUP have been designed and tested [3, 15, 16, 34, 36]. These implements have helped improve small-holder farmer access to markets, and they have improved the livelihoods for many of those in sub-Saharan Africa. The Well-Driver PUP is a further example of alternative use for this versatile vehicle, and it is displayed in **Figure 16**. The Well-Driver PUP could potentially reduce dependency on manual labor for driven well installation in developing countries, improve productivity, and keep laborers safer [3, 10, 15].

In locations where the number of drilling rigs are reduced, and labor is scarce, the Well-Driver PUP vehicle could be used as an instrument of economic development, providing micro-business development opportunities focused on well installation. Installing low-cost driven wells would improve the availability, quality, and accessibility of water in locations lacking sufficient water supplies. In order to be successful, the Well-Driver PUP operation and components must remain low-cost, easy to maintain, and based on locally accessible materials as much as possible. Although the efficacy of the effort might be diminished, it is intended that operation of this vehicle should not require formal training in well drilling or geology. In locations of appropriate water table depth, this vehicle could decrease both the wait time to install and the final cost of installed sanitary water wells [10].

3.2 Potential locations of utility for the well-driver PUP in the developing world

Global estimates indicate that 68% of the Earth's freshwater resources are locked-up in ice and glaciers, and 30% are found within the ground [38]. Groundwater is the portion of the total precipitation that soaks into the earth's crust and percolates downward into the porous spaces within the soil and rock, where it remains, or

Figure 16.
Well-Driver PUP vehicle in operational configuration with the hydraulic driving ram erected [37].

potentially, from where it finds its way out to the surface [3, 10]. Groundwater serves as the world's largest source of fresh water, and it plays a critical role in meeting the household needs of people around the world [39]. Groundwater is the primary source of the world's drinking water, and it supplies water for agricultural and industrial activities worldwide [40].

Through testing, Horn demonstrated that the Well-Driver PUP could achieve depths of up to 7.0 m [3, 10]. Although not yet demonstrated through formal experimentation, Horn analytically determined that the vehicle should be capable of driving to depths of up to 15 m without significant changes to the apparatus [10]. The depth to the water table varies throughout the world, but significant areas of the world have water within 15 m. On-going work is aimed at increasing the proven maximum achievable depth for water wells with this equipment [5].

Prior to the work of Horn [10], the design of the Well-Driver PUP was a project of several senior capstone teams in the Agricultural & Biological Engineering Department of Purdue University. The first capstone team mounted the post driver onto the PUP frame using a static three-point hitch lower arms [41]. This design pivots on the rear balls of the lower link arms, thereby allowing the driving ram to be rotated by a hydraulic cylinder serving as the upper link arm. This allows the post-driver vehicle to have a more distributed weight during transport operations [42]. This has the effect of moving the mechanical driver from a vertical operational orientation to an inclined transport position.

A second capstone team refined the Well-Driver PUP with additional safety shielding around the hydraulic pump, fixed hydraulic leaks, and made various additional vehicle improvements [43]. A support cradle was added to allow the post driver to stabilize and minimize bouncing during travel. A transmission lockout was also added to force the vehicle's transmission to remain in neutral, while operational at high engine speeds. This prevented the vehicle, if bumped by accident during operation, from jumping into gear [43]. Various additional improvements were made to the vehicle by Horn [10] before any initial driving efforts could be carried-out. Modifications included rebuilding the engine, improving the efficiency of the hydraulic system, and fabricating outriggers to address the stability and weight distribution issues of the vehicle during well driving operation. **Figure 17** portrays the Well-Driver PUP in the transport position. One of the four outriggers designed by Horn [10] is clearly visible on the right rear of the PUP vehicle. The driver support members are also highlighted in the photograph. A third capstone team added well pipe support grips [44], and a fourth team worked on vehicle repairs and supplemental ram weight [45].

3.3 Well components required for tube well driving

The driving action performed manually when installing hand-driven tube wells, has been mechanized hydraulically by the Well-Driver PUP. The components required to install a tube well stack utilizing the Well-Driver PUP are similar to that of a standard hand-driven well. These components include: the well point, the drive couplings, galvanized steel piping, and a well point drive cap. These parts were utilized by Horn [10] during his experimentation and have previously been shown in **Figures 8–11**. In addition to these components, Horn fabricated a specialized tube well installation drive sleeve appropriate for driving 2″ diameter pipe [10].

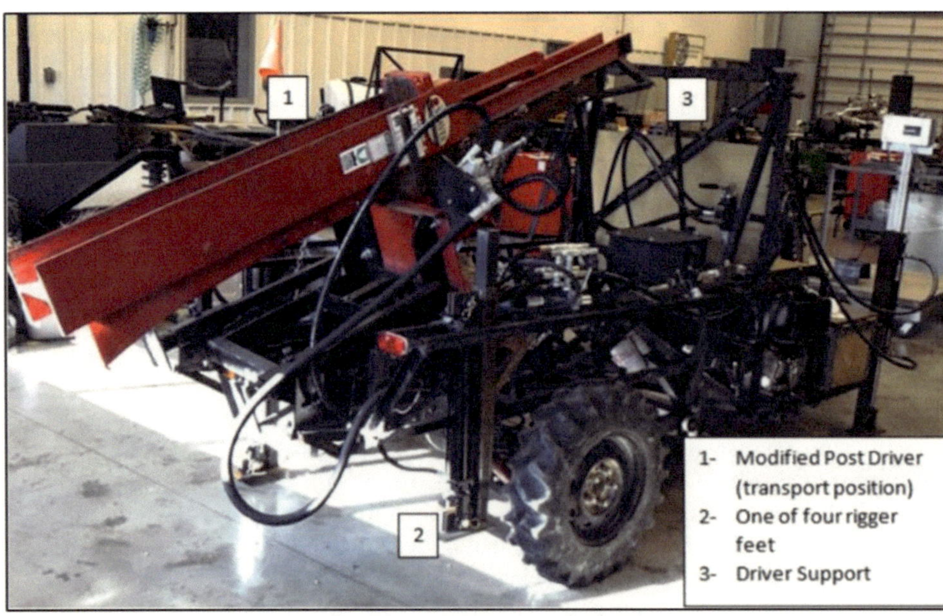

Figure 17.
Well-Driver PUP vehicle in transport configuration with the hydraulic driving ram collapsed into holding cradle [44].

Figure 18.
A comparison of the Horn 2″ tube well installation drive sleeve (left) and the Shaver (Graettinger, Iowa) steel fence post driving sleeve [10].

The Horn-manufactured well sleeve is shown on the left in **Figure 18**. The driving sleeve shown on the right in **Figure 18** is used for fence posts and was created by Shaver Manufacturing (Graettinger, Iowa) [10]. "The [well pipe] drive sleeve was constructed of 4–1/2″ × 3/16″ wall-drawn over-mandrel (DOM) steel tube, ½″ steel rod, and ½″ A36 steel plate" [10]. The DOM tubing was specifically chosen, because the internal weld bead is ground flush, as opposed to a raised weld bead, which would have posed alignment problems and caused unnecessary damage to the well point drive cap. The fabricated well point drive sleeve cleared with an approximated 0.6 cm gap around the diameter of a well point drive cap for a 2″ diameter pipe. Rub rails made from ½" steel rods were used for the inner steel channel and to position the center of the steel tube sleeve inline with the center of the driver strike plate. A ½″ steel plate was welded to the top of the drive sleeve to act as a "cap", serve as a wear plate, and to hold the sleeve in the proper position for striking [10]. The drive sleeve used by Horn [10] was designed for installing wells with a 2″ diameter pipe and is shown in position within the driving ram in **Figure 19**. The installation of a well with a different pipe diameter would require the fabrication of a new drive sleeve with similar features that matches the new desired well pipe diameter.

3.4 Key experience gaps to be addressed in well-driver well installations

Horn [10] hit water at some point during the driving process for the five test well installations that were completed. The results are shown in **Table 2**. These wells were

Figure 19.
Horn 2″ tube well installation drive sleeve positioned in hydraulic ram channel [10].

Well #	Well Depth (m)	Static Water Column in Well (m)	Confining Layers (m)	Water Supply
1	4.0	0.9	N/A	N/A
2	3.0	0.3	N/A	Intermittent
3	7.0	6.1	N/A	Continuous
4	5.2	N/A	2.7–3.4, 4.6–5.2	N/A
5	7.0	N/A	N/A	N/A

Table 2.
Summary of the Purdue Well-Driver PUP experience at tube well driving in Montgomery County, Indiana [3, 10].

all installed within Montgomery County, Indiana. The recovered water from the submersible pump varied from providing continuous flow, intermittent flow, or no flow [10]. Based on water supply ratings from these test installations, wells #1, #2, #4, and #5 were all deemed "dry" or "intermittent" wells. Well #3 provided a continuous water supply, and therefore, it was developed into a quality water well for testing purposes. This well had a depth of 7.0 m and a static water column of 6.1 m within the well [10].

The well pump utilized was a Waterra WSP-12 V-3B (Mississauga, Ontario) [46], which moved water at approximately 11.0 Lpm. Well #3 was completed, or finished, in accordance with the American Groundwater Trust procedure through the addition of a concrete pad, well surging, and disinfection [47]. The surging process removes the fine particles accumulated at the bottom of the well tube near the pump inlet, and it prevents these from being transferred into the drinking water drawn from the well [29, 48]. Surging also helps pack layers of fine particles around the well water inlet screen, which can then act as a "pre-filter" for large particles. The installation of a driven well of any other size than the 2″ diameter by the Well-Driver PUP would have required the purchase or fabrication of a wellpoint, drive cap, couplings, and galvanized steel piping, to match the diameter of the pipe desired. A larger pump would also have been required to surge and condition the larger diameter well for finishing.

3.5 Water quality results

Water quality samples were collected and submitted to the Montgomery County Health Department for analysis. The water quality parameters evaluated were Total Coliform and *E. Coli* count. Fecal Coliform count was not checked. These tests were done on a present/absent (P/A) basis. Installation of any new well in Indiana requires that upon receiving a continuous flow rating, a water quality sample be collected and sent to the perspective County Health Department for analysis. This was carried-out by Horn [10] in accordance with state regulations [29, 48]. The water quality test for Well #3 was reported to be absent for Total Coliform and *E. Coli* count, and therefore, satisfactory. Since the Montgomery County Health Department deemed the sample satisfactory, the water sample is considered "at the time of examination bacteriologically safe based-on U.S. EPA standards" [3, 10, 29, 48].

3.6 Potential impact

A high-resolution global-scale groundwater model was developed by de Graaf et al. [39]. A high-resolution global-scale groundwater model was created, highlighting the current water table depth on a global scale through computer model simulations. Most global-scale hydrological models (GHMs) do not include groundwater flow as a component of the model, due to the lack of consistent geohydrologic data available on a global scale. This model, run at 6° of resolution, utilized MODFLOW (U.S. Geological Survey, Washington, DC) to construct an equilibrium water table at its natural state. The aquifer schematization and properties used were based on the globally available data sets of lithology and transmissivities, combined with the thickness of an upper, unconfined aquifer. The model was initialized using outputs from the land-surface PCRaster Global Water Balance (PCR-GLOBWB) model (Utrecht University, Utrecht, Netherlands), which included net recharge and surface water levels. A sensitivity analysis of the various parameter settings was performed, and it showed that the greatest variation in saturated conductivity had the largest impact on estimates of the groundwater levels. The model validation with observed groundwater levels demonstrated that the predicted levels are reasonably well simulated for many regions of the world, particularly for sediment basins (R^2 = 0.95). These simulated regional-scale groundwater patterns help to provide insight into the availability of groundwater globally [39].

Figure 20 provides the expected water table depth below the land surface throughout the world based on the de Graaf et al. simulation [39]. Worldwide, there are many locations where the water table depth is projected to be within the 10–20 m

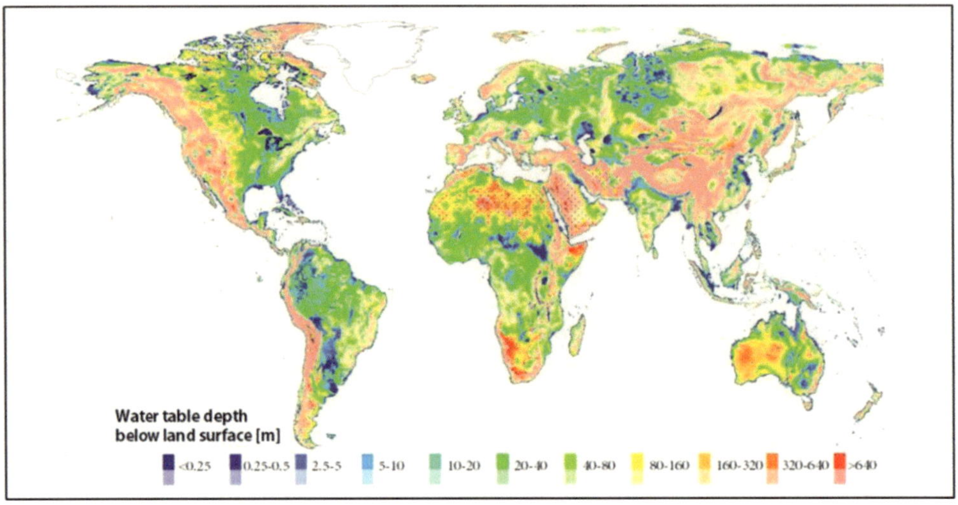

Figure 20.
Output for the estimated water table depth below land surface [m] from geological model [39].

range according to this model. Many of these locations are in sub-Saharan Africa, South America, northern India, Asia, and parts of the Asia Pacific Islands. These predictions simply identify potential locations where the Well-Driver PUP might be utilized, if sufficient depth can be demonstrated on a repeatable basis.

According to the United Nations, approximately 14.5% of the World's population is located within sub-Saharan Africa [1]. The Earth is projected to hold 9.8 billion people in 2050 and 11.2 billion people by 2100 [49]. The population of sub-Saharan Africa alone, is predicted to nearly double by 2050 [1]. Of the ten largest countries worldwide, Nigeria, which is located in sub-Saharan Africa, is growing the most rapidly [49]. It is projected to surpass the United States in population and become the third largest country in the world shortly before 2050 [49]. Clearly, when considering locations where the Well-Driver PUP could impact the largest number of people, consideration should be given to locations within sub-Saharan Africa.

The International Groundwater Resources Assessment Centre (IGRAC) and the UNESCO International Hydrological Programme have mapped-out the transboundary aquifers within Africa [50]. IGRAC, in collaboration with the British Geological Survey and the University College London, has developed maps to quantify the groundwater resources within Africa based upon local well data and known geological conditions. Their results for aquifer depth are highlighted in **Figure 21** and provide a high-resolution look at the depth to water table in Africa. When specifically looking at sub-Saharan Africa, the approximate depth to the groundwater is predominately less than 25 m, followed by areas in the 25–50 m range. These data demonstrate the vast number of locations within sub-Saharan Africa, where the Well-Driver PUP could possibly access groundwater. Once fully developed, this technology has great potential to have a substantial positive impact for the people in those regions of the World with little to no water access.

3.7 Potential people served per well estimates

The economic analysis of a water well installation depends upon the number of people that the well can serve, the depth of the water being pumped, the daily per capita water required, the duty cycle of the water pump, and the ability of the

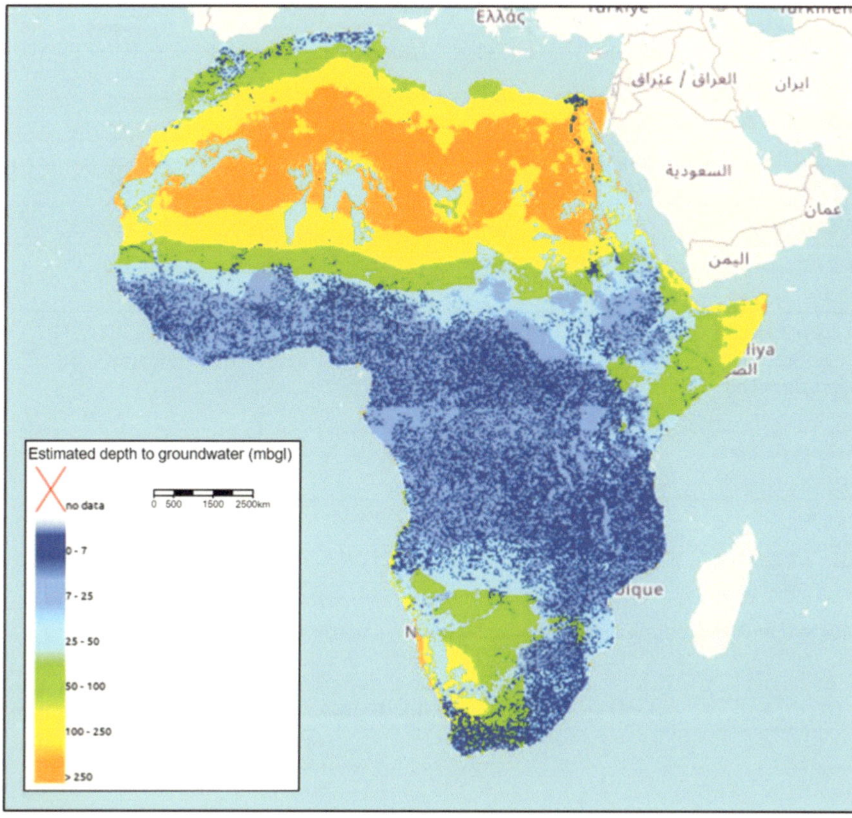

Figure 21.
Estimated depth to groundwater (mbgl) and transboundary aquifer of Africa based upon geologic and well drilling data [50].

water bearing formation to be sustained during pumping [10]. Additionally, the water table drawdown may affect the pumping flow rate over time. The successful water well installed by Horn [10] had a depth of 7 m and a pumping rate of approximately 11 L/m. If a continuous duty cycle on this well was run for 8 h per day, a total of 5400 L of water could be pumped [10]. According to the Sphere Humanitarian standards for Developing Countries [4] and WHO [8], the minimum water requirement is 15 L per person per day. Therefore, with a handpump, the maximum number of people using the Horn #3 well on a per day basis should not exceed 500, and a minimum flow rate of 16.6 L/m would be needed [4]. This guideline assumes that the water point is accessible for approximately 8 hours of the day [4]. Using these developed metrics, Horn's initial successful test well [10], could therefore supply water for between 270 and 360 people, if installed in a community setting.

3.8 Well cost per depth and value proposition estimates

Horn [10] conducted a cost analysis per well depth ranging from 1 to 35 m, using the U.S. market prices of the driven well components in 2019. This was updated to reflect current prices of those components in 2022, and the results are shown in **Table 3**. The percent difference between the pricing periods was calculated to determine how much change in the costs have occurred during the intervening years. Even with the recent spike in steel

Materials for Well	Price in 2019 (USD)	Price in 2022 (USD)	Percent Difference	Per	Source	Length (m)
2″×10′ Galvanized Steel Pipe	37.12	56.96	35%	each	HD	3
2″×36″ Well Point	60.66	66.89	9%	each	HD	1
2″ Pipe Coupling	12.86	11.63	−11%	each	HD	0
2″ Well Point Drive Cap	17.29	15.99	−8%	each	HD	N/A
2″×5′ Galvanized Steel Pipe Section (plus cut/threading cost)	20	20a	0%	each	HD (approximate)	1.5
Submersible Water Pump	Not Reported	224.10	0%	each	Waterra	N/A
1–80 lb. Bag of Concrete	Not Reported	5.87	0%	each	HD	N/A

Table 3.
Driven tube well material costs for various required installation components [10].

Well Depth (m)	Actual Length (m)	Well-Driver Cost in U.S. (USD)	Cost in Ghana (USD)
1.5	2.4	512	565
3.0	4.0	530	687
4.6	5.5	549	809
6.1	7.0	567	931
7.6	8.5	585	1053
9.1	10.1	604	1175
10.7	11.6	622	1297
12.2	13.1	640	1419
13.7	14.6	659	1541
15.2	16.2	677	1663
16.8	17.7	695	1785
18.3	19.2	714	1907
19.8	20.7	732	2028
21.3	22.3	750	2150
22.9	23.8	769	2272

Table 4.
An updated driven tube well cost per depth projection for 2022 based upon US costs and a comparison against prices in Ghana [10].

prices, the net changes have been relatively small. **Table 4** provides an updated well cost per depth using the current prices in 2022, modeled after Horn's initial calculations. This table also compares tube well costs for installations in Ghana at the same equivalent depth. One meter is the length of the 2″ well point, and it was therefore selected as the minimum depth [10]. Economic assumptions included in the analysis were that the well casing extended 1 m below the ground surface and that any section of pipe used which was less than 1.5 m was considered to have the full cost of a 1.5 m section [10]. A maximum depth of 15 m was chosen, as recommended by the book *Groundwater and Wells* [14]. The text states that well points driven by hammers massing 113 to 454 kg (1.1–4.5 kN) should be able to reach depths of 15 m or more under favorable situations [14]. The effective weight of the spring powered driving ram, per the Shaver HD-8 Operator's Manual, is 1.6 kN, which indicates that a 15 m deep well should be within an acceptable potential cutoff range for the driver without any modifications [10, 51]. This cost per depth table is carried-out to below 15 m, in the event that such depths can be reached through further development.

If additional modifications can be made to allow the driver to attain still deeper depths, then **Table 4** could easily be expanded. The average cost of a drilled well in the United States without a well casing is reported to have a range of US$49–98/m, or potentially up to US$164/m in tough soil conditions [52]. By averaging the cost per depth from **Table 4**, a driven well within the capabilities of the driver would average about US$68.05/m in physical material costs, excluding the Well-Driver Pup vehicle cost and fuel costs.

Based upon these calculations, it is possible that a driven well could be cheaper per m, even in a U.S. context, than previously reported by Horn [10]. A tube well, driven by a Well-Driver PUP, could certainly be a cost-effective alternative to a drilled well in a developing country. The current average cost per depth in Ghana is US$80/m for a tube well, and US$109/m for a drilled well [53, 54]. These values were first cited by Namara et al. [53] in 2011, and in this work, they have been updated to account for the depreciation of the Ghana Cedi between 2011 to 2022. The price increases for steel have nearly doubled its total cost since Horn study [10] in 2019, clearly necessitating a reevaluation of the cost structure for tube well installation. The calculations in **Table 4** account for pump, concrete, and labor costs. According to the U.S. Bureau of Labor Statistics (2022) [55], the average hourly wage for labor within the Water, Sewage, and Other Systems industry is $34.86 per hour [55]. In Ghana, the hourly rate is $2.76 USD per hour [56]. It is assumed that 6 hours of labor would be required for a well installation. Fuel costs are excluded from these calculations, due to variation of fuel pricing in both the U.S. and Ghana. This indicates that using the Well-Driver PUP to install a driven well in Ghana would be on average a potential savings of 49% compared to the current process. There are numerous locations throughout the world, and more specifically in sub-Saharan Africa, where the depth to groundwater is within the Well-Driver PUP's potential depth of 15 m [39, 50]. However, this proposition requires further testing to prove that that this depth can be repeatably achieved, since the Horn 2019 study [10] only demonstrated an experimental depth of 7 m. Furthermore, increasing the achievable driving depth capabilities of the vehicle through design improvements would increase the number of locations across the globe where the vehicle could provide improved access to groundwater.

Over the last 8 years, the primary author has had significant professional experience working in Ghana on various international development projects related to agriculture and water, sanitation, and hygiene (WASH) programming and continues to have access to a variety of networks within the country that are expressing interest in the use of the Well-Driver PUP, once it becomes proven and commercially viable. Therefore, consideration was initially given to the appropriateness of the Well-Driver PUP, if tested and eventually available within Ghana. Located in sub-Saharan Africa, Ghana has a long history of accessing shallow groundwater for the purpose of agricultural irrigation [57]. This experience has predominately been in the Keta Strip, within the Volta Region. Groundwater has been accessed through various power sources to lift water, including human feet/hand-operated equipment, such as rope and buckets. Access to many traditional shallow groundwater sources was extremely labor-inefficient, often being via hand-dug wells. Newer systems exist, but they are often out of reach for an individual farmer or household, due to the capital costs of acquisition. Currently, there are a total of 34,263 wells recorded within the Keta District. However, these are predominantly used for irrigation purposes. The small depth range of current tube wells within this area is between 6 to 9 m, and water is lifted primarily through small electrically powered pumps, located near the tube well. There is increasing potential within this area, due to the shallow alluvial depths of the aquifers being less than 20 m below the surface throughout the dry season [57].

Ghana's precipitation, surface water, and largely untapped groundwater resources are wholly sufficient to meet most of their projected water needs [58]. Ghana's groundwater resources are predominantly untapped with ample room for scale-up, particularly for agricultural purposes [59]. Ghana's groundwater aquifers range from between 10 to 60 m in depth, with well yields rarely exceeding 6 m^3/h. However, these wells yields can be much higher and the depths can be much deeper in areas where limestone is present within the soil profile [60]. As highlighted by de Graaf, et al. [39] and IGRAC [50], portions of Ghana's groundwater are accessible within 15 m of the surface. In areas where the water table is within Horn's previously demonstrated depth, such as portions of the Keta Strip, the potential to benefit individuals already exists at 7 m. However, if depths of up to 15 m can be achieved, a far greater number of locations could also benefit from this well installation process. As per **Table 4**, a 49% cost savings to install shallow tube wells in Ghana by using the Well-Driver PUP technology is extremely promising.

4. Conclusions

In developing countries, water is not always palatable and available in sufficient quantities. In many locations around the world, people lack sufficient access to water for both drinking and domestic purposes, and they use unsafe water sources. Water-related diseases pose a major risk to individuals through the consumption and unsafe use of poor water quality sources [4]. This is particularly true in sub-Saharan Africa. People must have equitable and affordable access to safe and sufficient water that is potable and in sufficient quantity for both drinking and domestic purposes. Stored underground water that can be removed by wells is the most likely means to supply this need. In many countries around the globe, individuals obtain their drinking water from community wells, so this kind of water access is commonly used for drinking and domestic purposes.

Worldwide, there are many locations where the water table depth is less than 15 m, specifically in the 10–20 m range. Many of these locations are within sub-Saharan Africa. Ghana is one of the many countries located within sub-Saharan Africa where the Well-Driver PUP could have a positive impact on the quality of life for those living there. Horn [10] installed a series of test wells, with the deepest being that of 7.0 m. This well received a continuous water rating and was formally completed [10]. The well water quality results analyzed received a satisfactory rating from the Montgomery County health authorities [3, 10, 29, 48]. Horn's initial test well could potentially supply water for between 270 and 360 people, if installed in a similar community setting.

This chapter reviewed the three primary types of wells utilized to obtain groundwater resources: dug, drilled, and driven wells [11]. Water wells of these three types can be installed through either manual or powered methods [11]. The Well-Driver PUP is a low-volume manufactured utility vehicle with a hydraulic post driver mated to it, to mechanize tube well installation. The components required to install a driven tube well stack utilizing the Well-Driver PUP include: the well point, the drive couplings, galvanized steel piping, a well point drive cap, and drive sleeve. The implementation and dissemination of the Well-Driver PUP technology has the potential to improve water access of safe water in developing countries for both drinking and domestic purposes.

Acknowledgements

The authors wish to thank the reviewers and journal editors for their gracious contribution of time and effort to improve this manuscript and ensure its publication. Dr. John Lumkes is recognized for his gracious donation of a PUP vehicle to further this research. Dr. Carol S. Stwalley is thanked for her editorial and proof-reading services in preparation of this manuscript. Elvis Kan-uge is thanked for his review and cultural insight. All cited authors that provided permissions for the reuse of their materials are hereby acknowledged and respectfully thanked. This research did not receive any specific grant resources of any kind from agencies in the public, commercial, or not-for-profit sectors. However, the assistance of the Purdue University Agricultural and Biological Engineering department is gratefully acknowledged for its support over the years with graduate teaching assistanceships and faculty salaries. Purdue University is an equal opportunity/equal access employer and service provider.

Conflict of interest

The authors declare no conflict of interest.

Author details

Grace L. Baldwin Kan-Uge, Tyler J. McPheron, Zackariah J. Horn and Robert M. Stwalley III*
Purdue University Agricultural and Biological Engineering, Indiana, USA

*Address all correspondence to: rms3@purdue.edu

References

[1] United Nations Department of Economic and Social Affairs, Population Division. World Population Prospects 2019: Highlights. UN, New York. 2019. [Online]. Available from: https://www.population.un.org/wpp/Publications/Files/wpp2019_10KeyFindings.pdf [Last Accessed 15 December 2020]

[2] Sweetman C, Medland L. Introduction: Gender and water, sanitation, and hygiene. Gender and Development. 2017;**25**(2):153-166. DOI: 10.1080/13552074.2017.1349867

[3] Horn ZJ, Stwalley RM III. Design and testing of a mechanized tube well installer. Groundwater for Sustainable Development. 2020;**11**:100442. DOI: 10.1016/j.gsd.2020.100442

[4] SPHERE. The SPHERE Handbook: humanitarian charter and minimum standards in humanitarian response, Geneva: SPHERE. 2018. [Online]. Available from: https://www.spherestandards.org/handbook-2018/ [Last Accessed 15 December 2020]

[5] Baldwin GL, Stwalley III RM. Improving the driving capabilities of a Well-Driver PUP (Purdue Utility Project) to install low-cost driven water wells. ASABE Annual International Meeting - Houston. 2022. DOI: 10.13031/aim.202200203

[6] Nolan B, Hitt K, Ruddy B. Probability of nitrate contamination of recently recharged ground waters in the conterminous United States. 2002. [Online]. Available from: https://www.water.usgs.gov/nawqa/nutrients/pubs/est_v36_no10/. [Last Accessed 15 December 2020]

[7] Weyer PJ, Cerhan JR, Kross BC, Hallberg GR, Kantamneni J, Breuer G, et al. Municiple drinking water nitrate level and cancer risk in older women: The Iowa Women's Health Study. Epidemiology. 2001;**12**(3):327-338. DOI: 10.1097/00001648-2,001,050,000-00013

[8] World Health Organization (WHO). What is the minimum quantity of water needed?. 2016. [Online]. Available from: https://www.who.int/water_sanitation_health/emergencies/qa/emergencies_qa5/en/. [Last Accessed 1 January 2019]

[9] UNICEF. Collecting water is often a colossal waste of time for women and girls. 2021. [Online]. Available from: https://www.unicef.org/press-releases/unicef-colelcting-water-often-colossal-waste-time-women-and-girls. [Last Accessed 16 December 2021]

[10] Horn ZJ. Prototyping a well-driver PUP (Purdue Utility Project) to install low-cost driven water wells. Purdue University [unpublished Master's thesis], West Lafayette, IN. 2019

[11] Proby F. Shallow well drilling. 2013. [Online]. Available from: https://www.lifewater.org/wp-content/uploads/2018/10/lifewater-Well-Drilling-Manual.pdf. [Last Accessed 11 October 2021]

[12] Wellowner. Types of wells. 2015. [Online]. [Last Accessed 17 November 2019]

[13] Carroll J. Personal communication. July, 14, 2019

[14] Driscoll FG. Groundwater and Wells. Saint Paul, MN: Johnson; 1986 LoC 85-63,577

[15] Horn ZJ, Stwalley III RM. Well-Driver PUP. ASABE Annual

International Meeting - Detroit. 2018. DOI: 10.13031/aim.201801196

[16] Horn ZJ, Stwalley III RM. A low-cost mechanized tube well installer. ASABE 2019 Annual International Meeting - Boston. 2019. DOI: 10.13031/aim.2019011319

[17] World Health Organization, UNICEF progress on sanitation and drinking water - 2013 update, WHO Press, Geneva. 2013. [Online]. Available from: https://www.data.unicef.org/resources/progress-on-sanitation-and-drinking-water-2013-update/ [Last Accessed 15 December 2020]

[18] Millennium Challenge Corporation. Measuring results of the Ghana water and sanitation sub-activity. 2017. [Online]. Available from: https://www.mcc.gov/resources/pub-pdf/report-ghana-closed-compact [Last Accessed 17 November 2019]

[19] Graham JP, Hirai M, Kim SS. An analysis of water collection labor among women and children in 24 subsaharan African countries. PLoS One. 2016;**11**(6):e0155981. DOI: 10.1371/journal.pone.0155981

[20] Mekonnen M, Hoekstra AY. Four billion people facing severe water scarcity. Science Advances. 2016;**2**(2):e1500323. DOI: 10.1126/aciadv.1500323

[21] Stevenson E, Ambelu A, Caruso B, Tesfaye Y, Freeman M. Community water improvement, household water insecurity, and women's psychological distress: an intervention and control study in Ethiopia. PLoS One. 2016;**11**(2):1-13. DOI: 10.1371/journal.pone.0153432

[22] Bisung E, Elliott SJ. Psychosocial impacts of the lack of access to water and sanitation in low- and middle-income countries: A scoping review. Journal of Water and Health. 2017;**15**(1):17-30. DOI: 10.2166/wh.2016.158

[23] Routray P, Torondel B, Clasen T, Schmidt WP. Women's role in sanitation decision making in rural coastal Odisha, India. PLoS One. 2017;**12**(5):e0178042. DOI: 10.1371/journal.pone.0178042

[24] Mahama AM, Anaman KA, Osei-Akoto I. Factors influencing householders' access to improved water in low income urban areas of Accra, Ghana. Journal of Water and Health. 2014;**12**(2):318-331. DOI: 10.2166/wh2014.149

[25] Adams E, Boateng A, Amoyaw G. Socioeconomic and demographic predictors of potable water and sanitation access in Ghana. Social Indicators Research. 2016;**126**(2):673-687. DOI: 10.1007/s11205-015-0912-y

[26] Sheuya SA. Improving the health and lives of people living in slums. Annals of the New York Academy of Sciences. 2008;**1136**(1):298-306. DOI: 10.1196/annals.1425.003

[27] World Water Assessment Program. The United Nations world water development report 2015: Water for a sustainable world, facts and figures, United Nations, New York. 2015. [Online]. Available from: https://www.unwater.org/publications/un-world-water-development-report-2015 [Last Accessed 15 December 2020]

[28] Leahy C, Winterford K, Nghiem T, Keheller J, Leong L, Willetts J. Transforming gender relations through water, sanitation, and hygiene programming and monitoring in Vietnam. Gender and Development. 2017;**25**(2):283-251. DOI: 10.1080/13552074.2017.1331530

[29] Professions and Occupations: Water Well Drilling. Indiana Code § 25-39-3. 2011

[30] H2O for Life, "Borehole well,". 2019. [Online]. Available from: https://www.h2oforlifeschools.org/page/borehole-well. [Last Accessed 17 November 2019]

[31] Koegel RG. Small diameter wells. 1985. [Online]. Available from: https://www.fao.org/3/X5567E/x5567e00.htm#Contents. [Last Accessed 15 December 2021]

[32] U.S. Army. US Army Field Manual: Well Drilling Operations (FM5-484). US Department of Defense, Washington, D.C. 1994

[33] Home Depot. Southland 3/4 in x 10 ft. Galvanized Steel Pipe-564-1200HC. 2021. [Online]. Available from: https://www.homedepot.com/p/Southland-3-4-in-x-10-ft-Galvanized-Steel-Pipe-564-1200HC/100534625. [Last Accessed 27 October 2021]

[34] Robison JP. Transportation and power solutions for Africa: The assessment and optimization of the Purdue Utility Platform. Purdue University [unpublished Master's thesis], West Lafayette, IN. 2016

[35] Purdue College of Engineering. Purdue Utility Project, Purdue University, West Lafayette, IN. 2019. [Online]. Available from: https://www.engineering.purdue.edu/pup/ [Last Accessed 15 December 2020]

[36] Wilson DD. Investigation of an affordable multigrain thresher for smallholder farmers in Sub-Saharan Africa 2015. Purdue University [unpublished Master's thesis], West Lafayette, IN

[37] Baldwin GL, Stwalley RM III. Opportunities for the scale-up of irrigation systems in Ghana, West Africa. Sustainability. 2022;**14**:8716. DOI: 10.3390/su14148716

[38] United States Geological Service. How much water is there on, in, and above the Earth?. 2016. [Online]. Available from: https://water.usgs.gov/edu/earthhowmuch.html. [Last Accessed 30 September 2021]

[39] de Graff IEM, Sutanudjada EH, van Beek LPH, Biekens MFP. A high-resolution global-scale groundwater model. Hydrology and Earth System Sciences. 2015;**19**(2):823-837. DOI: 10.5194/hess-19-823-2015

[40] Wada Y, Wisser D, Bierkens MFP. Global modeling of withdrawal, allocation, and consumptive use of surface water and groundwater resources. Earth Systems Dynamics. 2014;**5**:15-40. DOI: 10.5194/esd-5-15-2014

[41] Zeller C, Vaughn W, Well driver. Purdue University [unpublished Capstone report]. IN: West Lafayette; 2016

[42] Zhao S. Well driver PUP vehicle weight distribution during transport and operation. Purdue University [unpublished Masters thesis], West Lafayette, IN. 2022

[43] Chen Y, Jarufe J, Yun C. Renovations to the Purdue utility well driver. Purdue University [unpublished Capstone report], West Lafayette, IN. 2016

[44] Adams J, Hampston T, Wilson H. Improvements to the Purdue Utility Platform Well-Driver. Purdue University [unpublished Capstone report], West Lafayette, IN. 2019

[45] Fidler M, McPheron T, Soloman R. PUP Well Driver. Purdue University [unpublished Capstone report], West Lafayette, IN. 2022

[46] Waterra. Submersible Groundwater Sampling Pumps: 40, 60, or 90 foot lift.

2022. [Online]. Available from: https://waterra.com/product/submersible-groundwater-sampling-pump/. [Last Accessed 19 November 2022]

[47] American Ground Water Trust. agwt.org, 2012. [Online]. Available from: https://www.agwt.org/content/water-well-disinfection-procedure. [Last Accessed 15 December 2021]

[48] Water Well Drillers and Well Water Pump Installers. 312 Indiana Administrative Code § 13-1 – 13-12. 2015

[49] Held A. U.N. says world population will reach 9.8 billion by 2050. National Public Radio. 2017. [Online]. Available from: https://www.npr.org/sections/thetwo-way/2017/06/22/533935054/u-n-says-world-s-population-will-reach-9-8-billion-by-2050 [Last Accessed 15 December 2020]

[50] IGRAC. Africa groundwater portal. 2017. [Online]. Available from: https://www.un-igrac.org/special-project/africa-groundwater-portal. [Last Accessed 15 December 2021]

[51] Shaver Manufacturing, Operator's Manual for Hydraulic Post Driver Model HD-8 & HD-8-H, Graettinger, IA. 2009. [Online]. Available from: http://www.shavermfg.com/media/uploads/HD8-Operator-Manual.pdf [Last Accessed 15 December 2020]

[52] Home Advisor. How much does it cost to drill or dig a well?. 2021. [Online]. Available from: https://www.homeadvisor.com/cost/landscape/drill-a-well/. [Last Accessed 27 October 2021]

[53] Namara RE, Horowitz L, Nyamadi B, Barry B. Irrigation developments in Ghana: Past experiences, emerging opportunities, and future directions. International Food Policy Research Institute (IFPRI), Washington, DC. 2011. [Online]. Available from: https://www.reliefweb.int/report/ghana/irrigation-development-ghana-past-experiences-emerging-opportunities-and-future#:~:text=Ghana%20is%20endowed%20with%20sufficient,%3B%20Agodzo%20and%20Bobobee%201,994 [Last Accessed 15 December 2020]

[54] Kan-uge E. Personal communication. May, 20 2022

[55] U.S. Bureau of Labor Statistics. Occupational Employment and Wage Statistics. 2022. [Online]. Available from: https://www.bls.gov/oes/current/oes537072.htm#(2). [Last Accessed 19 November 2022]

[56] Yawson K. Personal communication. November, 7, 2022

[57] Namara RE, Horowitz L, Kolavalli S, Kranjac-Berisavljevic G, Dawuni BN, Barry B, et al. Typology of Irrigation Systems in Ghana. Columbo, Sri Lanka: International Water Management Institute (IWMI); 2010. DOI: 10.5337/2011.200

[58] Baldwin GL, Stwalley RM III. A review of freshwater programming and access options in Ghana, West Africa. African Journal of Water Conservation and Sustainability. 2022;**10**(4):1-19 LoC: 2375-0936

[59] Baldwin GL, Stwalley III RM. An economic analysis: the scale-up of irrigation systems in Ghana, West Africa. ASABE Annual International Meeting - Pasadena. 2021. DOI: 10.13031/aim.212100009

[60] Ghanaian Water Resources Commission (WRC), Water Use. 2020. [Online]. Available from: https://wrc-gh.org/water-resources-management-and-governance/water-use/. [Last Accessed 11 December 2020]

Chapter 4

The Importance of Salty Groundwater in the Supply of Thalassotherapy: The Case of Portugal

Luís Manuel Ferreira-Gomes, Luís José Andrade Pais and Pedro Jorge Coelho Ferreira

Abstract

Thalassotherapy is a balneotherapy activity with a wellness and therapeutic character, using salty water captured at sea. That activity is very similar to thermalism, which uses natural mineral water (NMW) captured in aquifer systems. In Portugal, thermalism is a well-established activity, and for a medical spa to be in operation, there is a legal requirement to have two specific professionals: a medical doctor—Clinical Director—and a hydrogeologist—Technical Director (TD). The exploitation of the NMW is the responsibility of the TD, a professional with know-how in the field of hydrogeology. Thus, this chapter presents an introduction with some fundamental concepts about thermalism and thalassotherapy, generic aspects of those activities in Portugal, and their objectives. Methodological elements are presented, followed by the main results and interpretations, with the physic-chemical characteristics of Portuguese NMWs; of very salty special groundwaters, not licensed as NMW; and sea and ocean waters. Finally, the main conclusions and several considerations are presented, in the sense that the practice of exploiting salty groundwater, in the proximity of the sea, will be a new field for the activity of hydrogeological professionals, provided that they are classified as NMW, to use in the thermalism activity.

Keywords: natural mineral water, thermalism, seawater, salty groundwater, thalassotherapy

1. Introduction

Portugal, a country in the SW of Europe (**Figure 1**), has a great tradition in thermalism, which has distant origins in time and predates the Portuguese nation, with many places using special groundwater, namely when they were naturally hot, since the time of the occupation of this territory by the Roman people, about 2000 years ago.

Thermalism, in Portugal, corresponds to the use of natural mineral water (NMW) with balneotherapy techniques and/or treatments and other specialized treatments,

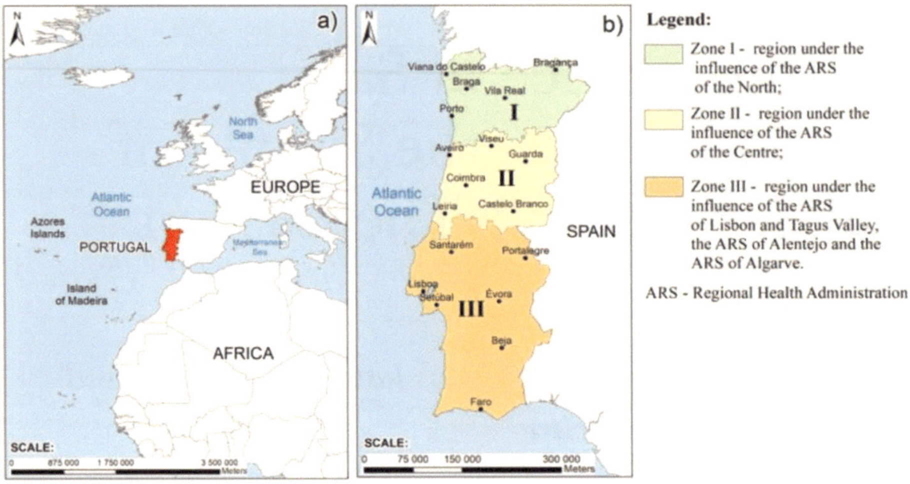

Figure 1.
(a) Geographical setting of Portugal (from ref. [1]); and (b) distinction of three regions under study in the organization of the AMN (from ref. [2]).

such as those used in the respiratory tract, among others, for therapeutic and wellness purposes. The official definition of thermalism, published in the medical spa law [3], is as follows: "Thermalism", corresponds to the use of NMW and other complementary means for prevention, therapy, rehabilitation, or wellness. A notion of the reality of the Portuguese thermalism can be obtained by consulting the Internet site of the Portuguese Medical Spa Association [4]. It should be noted that there are about 50 medical spas in operation in a normal year.

The NMW corresponds to groundwater, which, in order to have that designation, has to be classified by Portuguese law (Decree Law 86/1990 [5] and 54/2015 [6]), which guide the main procedures and studies to be developed, namely of the hydrogeological nature and of the quality of the groundwater under study, for 12 consecutive months, with physical–chemical and microbiological analyses of water samples collected at the head of the abstraction. Examples of studies in this sense can be seen in several published works [7–10]. NMW has characteristics that distinguish it from other groundwaters, such as the stability levels of their physical–chemical parameters [11]. This specific type of resource does not have recommended or admissible limits for the vast majority of physic-chemical parameters; only for some constituents of NMW (mostly trace chemical elements) concentration limits are established, in accordance with Directive 2003/40/EC, of 16 May [12].

Thalassotherapy in Portugal, generically, corresponds to the use of seawater, in hydrotherapy practices and the like, in a similar way to what is classic in thermalism. The term "Thalassotherapy" comes from the Greek [13]: "*thálassa*- sea and *therapeia*—treatment," being a neologism proposed by La Bonnardiére in 1867 to designate the therapeutic use of seawater, preferably in the form of baths and also taking into account that the treatments are normally carried out close to the sea so as to have a simultaneous action of the maritime climate.

In Portugal, unfortunately, there is no specific legislation in the field of thalassotherapy [14, 15]. Despite this situation, Portugal has some cases of excellence, with extraction of salty water from the sea; the existence of autonomous spas, that is, not included in hotels, deserves mention: (i) the Thalasso Costa da Caparica [16], with group hydrotherapy, massages, physiotherapy, esthetic services, and providing of medical consultations for advice and (ii) Barra Talasso Nazaré [17], with a dynamic swimming pool incorporating several techniques, such as hydromassage, several types of showers, walking circuit, individual hydromassage-application cabins, Vichy shower, Scottish shower, and others. Other spas included in hotels or hotel resorts, almost exclusively for leisure and wellness purposes, also exist in Portugal, such as in the Vilalara Thalasso Resort [18] and the Grande Real Hotel & Spa [19].

Some detailed elements on thalassotherapy in Portugal can be observed in several works [14, 15, 20, 21], where it is evidenced that in thalassotherapy, besides seawater being the basis of such activity, other elements related to the sea can be important in the global activity such as mud, seaweed, sand, and local climatic conditions.

The basic difference between a thalassotherapy center and a medical spa in thermalism is in the base fluid; thalassotherapy uses natural salt water collected from the sea, while a medical spa uses NMW, collected from an aquifer system of a geological formation.

In Portugal, thermalism is a well-established activity. For a medical spa to be operational, there is a legal requirement to have two specific professionals: a doctor—Clinical Director (CD)—and a hydrogeologist—Technical Director (TD). The exploration of NMW is the responsibility of the TD, a professional with know-how in the field of hydrogeology, who makes the NMW available from the aquifer system, at the head of the abstraction, to be used in the medical spa. Thalassotherapy, currently using water captured from the sea, not having the same rules as NMW in the exploration, is also not obliged to comply with the stability in quality that is required for NMW. In Portugal, thalassotherapy is not currently integrated in the thermalism sector, the latter being supervised by the Directorate-General for Energy and Geology (DGEG).

Still, within the scope of this chapter, it is important to emphasize that in these new and current times, in certain regions of the world, such as Portugal, with the current trends in climate change, NMWs tend to decrease their natural availability and, consequently, their production. If the classic hydrological cycle does not occur, with the current trends of lower precipitation and higher evaporation, this will put the sustainable exploitation of aquifer systems at risk, necessarily leading to reductions in exploitation flows and causing potential changes in the quality of the resource. So, as some of the aquifer systems that produce the NMWs are clearly in danger, new alternatives must be sought, and the use of salty groundwater, captured from free (unconfined) aquifer systems, close to the sea, will not pose a problem at least in terms of decreased flows over time, if they are exploited by hydrogeologists who comply with the good rules of the area, namely, being totally rigorous in not causing the advancement of the salt wedge intrusion into the interior of the continent.

Thus, in Portugal, there is a lot of knowledge and experience about thermalism. On the other hand, it is a European country, with an extensive Atlantic coastline, 832 km in continental Portugal, in addition to the 960 km of the archipelagos of Madeira and the Azores [22]. There are therefore conditions, especially in areas where sedimentary geological units occur on the coastline, which facilitate the penetration of the salt wedge to the interior. Based on hydrogeological studies, it will not be difficult to implement abstractions that allow to obtain a resource, natural salt water, with excellent stability in its chemical composition. Therefore, the authors' understanding is that there are conditions for, in many situations, underground abstraction of saltwater, often in free aquifers (unconfined) or other systems, for extracting water from the earthy massif, coming indirectly from the sea, but with the potential, in certain places, to maintain stability, because the passage of water through the aquifer system ends up purifying certain impurities, and thus, there is potential for physicochemical stability, as happens in natural mineral waters.

Those notions of the previous paragraph were the basis of academic work, essentially around architecture [22], designing a special thalassotherapy spa, for the Cabo Espichel, in Portugal, with this place presenting excellent characteristic potentiators of such a situation. Some elements of that project were presented in several papers [23, 24], presenting in **Figure 2** the scheme of principle for the abstraction of salt water, idealized near the sea.

The sketch of the abstraction in **Figure 2** is the result of an initial mechanical excavation with the aid of a boom-slew excavator, in order to carry out an excavation in the natural massif, which constitutes the unconfined aquifer. Three tubes are installed vertically, the two outer tubes made of rings of perforated concrete and the interior of stainless steel, grooved in its lower part, to allow the entry of salty water. Between those tubes, there is selected granular material, the outermost with coarse sand and the interior with coarse gravel, so that the underground flow evolves horizontally so as to gradually increase its velocity, until it enters the pumping chamber, inside the stainless-steel tube. In this way, the underground flow regime tends to remain laminar, enhancing not only the proper functioning of the entire system over time but also the quality of the water to be captured, as the filtering system works in a more effective, in case there are fine particles suspended in the water and even pathogenic bacteria. In the abstraction area, in order to minimize the potential for fluids to flow downwards vertically (rainwater and others), in the area most sensitive to contamination, an impermeable geomaterial is installed horizontally, which may be

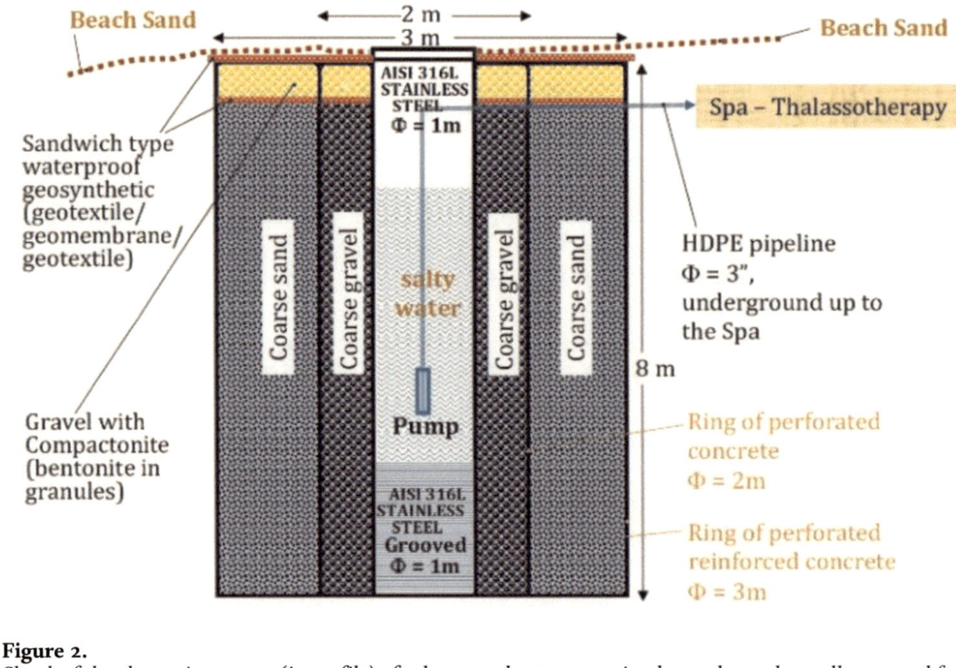

Figure 2.
Sketch of the abstraction system (in profile) of salty groundwater, as a circular and annular well, proposed for Lagosteiros Beach, to supply the thalassotherapy spa proposed for Cabo Espichel (from ref. [24]).

of the sandwich type, consisting of three levels, "geotextile/geomembrane/geotextile," not only with the mission of fulfilling the water-tightness but also to serve as a "separator," so that there is no mixing of materials from top to bottom. The strengthening of the water-tightness is provided by the 30 cm layer of compactonite (bentonite in granules). Finally, the whole set is headed with another geomaterial similar to the one already mentioned, in particular to serve as a separator for the local natural materials, so that the abstraction site has no visual impact. The adduction system to the bathhouse, also for reasons of not having a visual impact, must follow in a buried conduit. Regarding the environmental impact of this work, it exists only in the construction phase.

Therefore, the main objective of this chapter is to show that in Portugal, there is already a vast knowledge about thermalism; to present in particular the various types of special groundwater (NMW) that support this activity; to highlight the groundwater that has similarities with seawater; to show in the same language (using graphics and in particular Piper and Stiff diagrams) the main types of seawater (based on elements from the literature); and, finally, to make several considerations in the sense that in the future, thalassotherapy may be supplied by salty groundwater, captured in the proximity of the sea, so that this activity will have the potential for greater stability in its quality, potentially integrating itself in the NMW.

2. Methodological aspects

The present work is organized in two phases:

1st phase—establishment of an inventory of special groundwater occurrences in Portugal and of some ocean and sea waters, with physical–chemical results available in the literature. This phase is limited to the bibliographical research of published works with results of physical–chemical analyses of Portuguese NMW, as well as of other special groundwaters, some of which, in the past, were already classified as NMW but that in the meantime, over time, for various reasons, bureaucratically lost the title but still have emergences (springs) of quality, generically associated to old bathhouses, some in very advanced ruin; there are also some occurrences (springs) that have never been classified but have the potential to be classified. In some of these cases, these waters served therapeutic actions by popular movements but were not integrated within the official rules of thermalism. This phase also included research on sea and ocean waters from various regions of the world, including the results of physical and chemical analyses of their waters. The analytical methods used to perform the physical-chemical analyses are often described in the researched works; however, it is mentioned that they generally follow the standardized techniques and procedures [25, 26].

2nd phase—treatment and interpretation of data from physical–chemical analyses of the studied waters, in order to highlight the chemical types of the occurrences and their classifications. The main classifications used were on the total mineralization—M_T (**Table 1**), the pH, and the main ionic species in Piper and Stiff diagrams, using adequate software [28].

Besides the classifications presented, there are others, namely those that have to do with some singularities that the groundwaters present, these being the ones that are generally known in the field of thermalism in Portugal. Thus, in that sense, in this chapter, we follow what a recent publication [21] presented, which is in line with Directive 80/778/EEC [25], considering that water acquires its own identity, even with small amounts of certain chemical elements, and therefore, the same author [21] understands that it may also apply to NMW, resulting in the following designations:

- *sulfurous* waters, when titratable Sulfur (H_2S, HS^-, S^{2-}) ≥ 1 mg/L;
- *fluoridated* waters—fluorine (F^-) > 1.0 mg/L,
- *ferruginous* waters—bivalent iron (Fe^{2+}) > 1 mg/L
- *silicate* waters—colloidal free silica (SiO_2) > 10 mg/L,
- *carbonaceous* or *acidulated* (*gas-carbonate*) waters—free carbon dioxide (CO_2) > 250 mg/L (content at emergence).

The waters may also have other designations if any chemical element occurs singularly and exceptionally.

Designation	Total Mineralization—M_T (mg/L)
Hyposaline	< 200
Weakly mineralized	200–1000
Mesosaline	1000–2000
Hypersaline	> 2000

Table 1.
Classification of waters in relation to total mineralization [27].

3. Results and interpretations

3.1 Physical: Chemical aspects of Portuguese NMW

Portugal is a small country but with many and a great variety of special groundwaters, situations that result mainly from the territory being associated with many varieties of geological units. The oldest inventory of the main occurrences of natural springs, which deserves reference, was made by Fonseca Enriques in 1726 [29]. Most of the places where the main medical spas are currently located in Portugal, with the practice of thermalism, were already mentioned in that inventory.

On the chemical composition of Portuguese groundwater, it is worth mentioning the works carried out by Almeida and Almeida, from 1966 to 1988 [30–33], with the publication of several books for different regions, starting with the Algarve [30], in the south of Portugal; then the region of Trás-Os-Montes e Alto Douro [31]; the region of Beira Alta [32]; and, finally, the region of Minho [33], in the north. Those works present the results of physical–chemical analyses of the NMW that served the medical spas in operation at the time, as well as other waters, corresponding to older NMW, however, deactivated, and still, others, which had never had the NMW classification but with the potential to be. The results of the physical–chemical analyses, obtained at that time, are very much convergent with the results that are still obtained today with more developed physical–chemical research techniques. About the chemical composition of the Portuguese NMW, it is also worth mentioning the work of the DGGM [34], in 1992, with the presentation of detailed physical–chemical analyses of all the medical spas in operation at that time. Mention should be made of the Geothermal Resources Catalog, published by the Geological and Mining Institute, in 1999 [35], with the inclusion of results of physical–chemical analyses of occurrences with geothermal potential, as they have emergence temperatures above 20°C; many of the occurrences in this work also correspond to NMW in use in thermalism, and others, naturally, have the potential to obtain this classification. Finally, it is mentioned that

Place	T	pH	M_T	S_T	SiO_2	CO_2	Main ionic components (mg/L)	D	Ref.
Amarante	23	9.5	226	20	36	—	F^- = 24, HCO_3^- = 48, Cl^- = 20, SO_4^{2-} = 7, Na + =67, Ca^{2+} = 2, K^+ = 0.4	S	[37]
Aregos	62	9.2	313	22	54	—	F^- = 19, HCO_3^- = 85, Cl^- = 30, SO_4^{2-} = 9, Na^+ = 84, Ca^{2+} = 3, K^+ = 2	S	[34]
Caldas da Saúde	27	8.8	564	91	94	—	F^- = 18,$HCO_{3\_}$ = 76,Cl^- = 130,SO_4^{2-} = 48, Na + =158, Ca^{2+} = 6, K^+ = 7	S	[34]
Caldelas	33	8.3	136	1	23	—	F^- = 2, HCO_3^- = 61, Cl^- = 7, SO_4^{2-} = 11, Na^+ = 12, Ca^{2+} = 19, K^+ = 1, Mg^{2+} = 1	B	[34]
Cana-veses	30	9.4	290	36	39	—	F^- = 24, HCO_3^- = 70, Cl^- = 23, SO_4^{2-} = 2, Na^+ = 84, Ca^{2+} = 2, K^+ = 1	S	[34]
Carlão	29	8.2	434	20	61		F^- = 16, HCO_3^- = 226, Cl^- = 13, SO_4^{2-} = 2, Na^+ = 103, Ca^{2+} = 4, K^+ = 4	S	[34]
Carva-lhelhos	20	7.9	249	0	39	—	F^- = 3, HCO_3^- = 137, Cl^- = 3, SO_4^{2-} = 6, Na^+ = 50, Ca^{2+} = 6, K^+ = 2, Mg^{2+} = 1	B	[34]
Chaves	68	6.9	2508	0	74	1150	F^- = 8, HCO_3^- = 1665,Cl^- = 41, SO_4^{2-} = 28, Na^+ = 588, Ca^{2+} = 24, K^+ = 71	G	[34]
Eirogo	24	8.8	491	51	65	—	F^- = 15, HCO_3^- = 96, Cl^- = 87, SO_4^{2-} = 61, Na^+ = 139, Ca^{2+} = 6, K^+ = 5	S	[34]
Entre-os-Rios	19	8.7	466	158	43	—	F^- = 20, HCO_3^- = 160, Cl^- = 62, SO_4^{2-} = 1, Na^+ = 137, Ca^{2+} = 3, K^+ = 3	S	[34]
Gerês	47	9.1	281	1	67	—	F^- = 14, HCO_3^- = 81, Cl^- = 15, SO_4^{2-} = 10, Na^+ = 67, Ca^{2+} = 3, K^+ = 2	B	[34]
Melgaço	16	6.1	1983	0	68	2310	HCO_3^- = 1425, Cl– = 14, SO_4^{2-} = 4, Na^+ = 128, Ca^{2+} = 275, Mg^{2+} = 50	G	[34]
Moimenta	30	9.0	462	4	32	—	F^- = 9, HCO_3^- = 71; Na^+ = 51	S	[21]
Moledo	45	9.2	242	22	36	—	F^- = 17, HCO_3^- = 71, Cl^- = 20, SO_4^{2-} = 4, Na^+ = 67, Ca^{2+} = 2, K^+ = 1	S	[35]
Monção	49	7.8	559	4	81	—	F^- = 12, HCO_3^- = 271, Cl^- = 42, SO_4^{2-} = 8, Na^+ = 125, Ca^{2+} = 10, K^+ = 6, Mg^{2+} = 3	S	[34]
Pedras Salgadas	17	6.2	2977	—	76	2550	HCO_3^- = 2078, Cl^- = 35, SO_4^{2-}=, Na^+ = 528,Ca^{2+} = 189,K^+ = 28,Mg^{2+} = 26	G	[34]
São Jorge	23	8.6	690	62	62	—	F^- = 17,HCO_3^- = 161,Cl^- = 202,SO_4^{2-} = 2, Na^+ = 209, Ca^{2+} = 5, K^+ = 9	S	[34]
São Lou-renço	30	8.0	360	2	51	—	F^- = 13, HCO_3^- = 185, Cl^- = 13, SO_4^{2-} = 4, Na^+ = 84, Ca^{2+} = 6, K^+ = 3	S	[35]
São Vicente	19	8.6	503	164	70	—	F^- = 21, HCO_3^- = 162, Cl^- = 61, SO_4^{2-} = 1, Na^+ = 137, Ca^{2+} = 4, K^+ = 5	S	[34]

Place	T	pH	M_T	S_T	SiO_2	CO_2	Main ionic components (mg/L)	D	Ref.
Taipas	31	8.2	256	8	36	—	F^- = 7, HCO_3^- = 90, Cl^- = 26, SO_4^{2-} = 26, Na^+ = 48, Ca^{2+} = 20, K^+ = 1	S	[34]
Terronha	30	8.8	376	7	27	—	F^- = 17, HCO_3^- = 162, Na^+ = 131	S	[21]
Vidago	18	6.5	5604	—	58	1960	F^- = 2, HCO_3^- = 3966, Cl^- = 66, SO_4^{2-} = 3, Na^+ = 1180,Ca^{2+} = 194,K^+ = 92,Mg^{2+} = 34	G	[34]
Vizela	50	9.3	345	30	54	—	F^- = 18, HCO_3^- = 83, Cl^- = 39, SO_4^{2-} = 12, Na^+ = 94, Ca^{2+} = 3, K^+ = 2	S	[32]

T—temperature at emergence (°C), M_T—total mineralization (mg/L), S_T-total sulfuration (in I_2 0.01 N—mL/L), SiO_2—colloidal free silica (mg/L), CO_2—free carbon dioxide (mg/L), D—typical designation as to chemical composition: B—bicarbonate, G—gas-carbonate, S—sulfurous.

Table 2.
Main physical-chemical parameters of special groundwater in zone I, classified as NMW, in use in thermalism.

Place	T	pH	M_T	S_T	SiO_2	CO_2	Main ionic components (mg/L)	D	Ref.
Águas	19	7.6	280	9	40	—	F^- = 3, HCO_3^- = 147, Cl^- = 12, SO_4^{2-} = 8, Na^+ = 65, Ca^{2+} = 1, K^+ = 1	S	[36]
Alcafache	48	8.4	360	1.6	52	—	F^- = 14, HCO_3^- = 154, Cl^- = 34, SO_4^{2-} = 3, Na^+ = 92, Ca^{2+} = 5, K^+ = 3	S	[34]
Almeida	31	8.3	527	7	30		F^- = 17, HCO_3^- = 261, Cl^- = 54, Na^+ = 148, Ca^2 $^+$ = 5, K^+ = 3	S	[38]
Azenha*	28	7.3	682	0	13	—	HCO_3^- = 239, Cl^- = 193, SO_4^{2-} = 30, Na^+ = 119, Ca^{2+} = 56, K^+ = 7, Mg^{2+} = 23	C	[34]
Bicanho	28	7.9	721	0	—	—	HCO_3^- = 243, Cl^- = 213, SO_4^{2-} = 34, Na^+ = 138, Ca^{2+} = 56, Mg^{2+} = 24	C	[35]
Carvalhal	23	9.3	336	25	56	—	F^- = 21, HCO_3^- = 88, Cl^- = 28, SO_4^{2-} = 5, CO_3^{2-} = 11, Na^+ = 92, Ca^{2+} = 2, K^+ = 2	S	[34]
Cavaca	28	8.3	360	9	56	—	F^- = 15, HCO_3^- = 164, Cl^- = 21, SO_4^{2-} = 2, CO_3^{2-} = 2, Na^+ = 87, Ca^{2+} = 6, K^+ = 3	S	[34]
Cró	23	8.2	395	19	48	—	F^- = 17, HCO_3^- = 165, Cl^- = 38, SO_4^{2-} = 11, Na^+ = 104, Ca^{2+} = 4, K^+ = 3	S	[35]
Curia	20	7.7	2375	—	10	—	F^- = 1, HCO_3^- = 217,Cl^- = 29, SO_4^{2-} = 1454, Na^+ = 20, Ca^{2+} = 591, K^+ = 2, Mg^{2+} = 51	Su	[34]
Felgueira	36	8.4	414	7	48	—	F^- = 16, HCO_3^- = 162, Cl^- = 5, SO_4^{2-} = 12, CO_3^{2-} = 3, Na^+ = 111, Ca^{2+} = 5, K^+ = 3, Li^+ = 1	S	[34]
Granjal*	23	9.1	507	47	56	—	F^- = 28, HCO_3^- = 149, Cl^- = 74, SO_4^{2-} = 14, CO_3^{2-} = 10, Na^+ = 157, Ca^{2+} = 2, K + =3	S	[38]
Longroiva	44	8.9	441	44	58	—	F^- = 23, HCO_3^- = 149, Cl^- = 44, SO_4^{2-} = 10, Na^+ = 122, Ca^{2+} = 3, K^+ = 5, Li^+ = 1	S	[35]
Luso	22	5.4	40	0	11	—	HCO_3^- = 8, Cl^- = 9, SO_4^{2-} = 2, NO_3^- = 1, Na^+ = 6, Ca^{2+} = 1, Mg^{2+} = 1, K^+ = 1	Si	[34]
Man-teigas	48	9.2	179	7	39	—	F^- = 10, HCO_3^- = 43, Cl^- = 6, SO_4^{2-} = 13, Na^+ = 43, Ca^{2+} = 3, K^+ = 1	S	[35]
Monfor-tinho	27	5.7	49	0	18	—	$HCO_{3\_}$ =15, Cl^- = 5, NO_3^- = 2, Na^+ = 5, Ca^{2+} = 1, Mg^{2+} = 1	Si	[34]

Place	T	pH	M_T	S_T	SiO_2	CO_2	Main ionic components (mg/L)	D	Ref.
Monte Real	18	7.2	2620	20	15	—	HCO_3^- = 126, SO_4^{2-} = 1575, Cl^- = 142, Na^+ = 83, Ca^{2+} = 626, Mg^{2+} = 48	Su	[34]
Sangemil	48	8.4	469	21	82	—	F^- = 17, HCO_3^- = 171, Cl^- = 54, SO_4^{2-} = 7, Na^+ = 114, Ca^{2+} = 6, K^+ = 5	S	[34]
São Miguel	19	5.9	78	0	25	—	HCO_3^- = 23, Cl^- = 7, NO_3^- = 6, Na^+ = 9, Ca^{2+} = 4, K^+ = 1	B	[8]
São Pedro do Sul	67	8.9	341	23	70	—	F^- = 17, HCO_3^- = 104, Cl^- = 29, SO_4^{2-} = 9, CO_3^{2-} = 6, Na^+ = 87, Ca^{2+} = 3, K^+ = 4	S	[34]
São Tiago	19	6.7	214	—	52	—	F^- = 1, HCO_3^- = 106, Cl^- = 9, SO_4^{2-} = 6, Na^+ = 17, Ca^{2+} = 14, Mg^{2+} = 8, Fe^{2+} = 3	B	[10]
Unhais da Serra	37	8.7	269	13	50	—	F^- = 15, HCO_3^- = 85, Cl^- = 25, SO_4^{2-} = 7, CO_3^{2-} = 3, Na^+ = 69, Ca^{2+} = 4, K^+ = 2	S	[35]
Vale da Mó	15	6.6	165	—	9	—	HCO_3^- = 89, Cl^- = 19, SO_4^{2-} = 9, Na^+ = 12, Mg^{2+} = 13, Fe^{2+} = 9	F	[32]

T—temperature at emergence (°C), M_T—total mineralization (mg/L), S_T—total sulfuration (in I_2 0.01 N—mL/L), SiO_2—colloidal free silica (mg/L), CO_2—free carbon dioxide (mg/L), D—typical designation as to chemical composition: B—bicarbonate, *C*—chloride, *F*—ferruginous, *S*—sulfurous, *Si*—silicate, *Su*—sulfate.
This site is not currently active.

Table 3.
Main physical-chemical parameters of special groundwater in zone II, classified as NMW, in use in thermalism.

Place	T	pH	M_T	S_T	SiO_2	CO_2	Main ionic components (mg/L)	D	Ref.
Caldas Rainha	35	6.9	2990	—	18	—	Cl^- = 1007, HCO_3^- = 315, SO_4^{2-} = 641, Na^+ = 660, Ca^{2+} = 270, Mg^{2+} = 57	C	[34]
Cucos	35	7.0	3393	—	18	—	F^- = 1, HCO_3^- = 378, Cl^- = 1584, SO_4^{2-} = 200, Na^+ = 1004, Ca^{2+} = 158, Mg^{2+} = 40, K^+ = 9	C	[34]
Enven-dos	22	4.7	26	0	10	—	HCO_3^- = 1, Cl^- = 7, SO_4^{2-} = 3, Na^+ = 4, Ca^{2+} = 1, Mg^{2+} = 1	Si	[34]
Estoril	35	7.1	5743	—	32	—	Cl^- = 2709, HCO_3^- = 285, SO_4^{2-} = 358, Na^+ = 1574, Ca^{2+} = 273, Mg^{2+} = 90, K^+ = 50	C	[35]
Piedade	24	7.0	2444	—	12	—	HCO_3^- = 393, Cl^- = 990, SO_4^{2-} = 207, Na^+ = 644, Ca^{2+} = 160, Mg^{2+} = 34	C	[34]
Sta Mta-Ericeira	17	6.8	3902	—	10	—	HCO_3^- = 314, Cl^- = 1395, NO_3^- = 500, Na^+ = 993, Ca^{2+} = 117, K^+ = 144	C	[21]
Vimeiro (Frades)	26	6.9	3300	—	12	—	HCO_3^- = 476, Cl^- = 1441, SO_4^{2-} = 229, Na^+ = 836, Ca^{2+} = 171, K^+ = 36, Mg^{2+} = 88	C	[34]
Cabeço de Vide	18	11.4	223	—	—	—	Cl^- = 60, OH^- = 30, CO_3^{2-} = 17, SO_4^{2-} = 4, Na^+ = 65, Ca^{2+} = 26, K^+ = 5	O	[34]
Castelo de Vide	15	7.2	1010	—	17	—	HCO_3^- = 343, Cl^- = 155, NO_3^- = 131, SO_4^{2-} =74, Na^+ = 108, Ca^{2+} = 78, K^+ = 58, Mg^{2+} = 44	N	[34]
Fadago. de Nisa	19	8.3	380	23	30	—	F^- = 10, HCO_3^- = 149, Cl^- = 71, SO_4^{2-} = 0.2, Na^+ = 100, Ca^{2+} = 11, K^+ = 2	S	[34]
Pizões (Moura)	18	7.0	736	—	24	—	HCO_3^- = 388, Cl^- = 102, SO_4^{2-} = 19, NO_3^- = 21, Na^+ = 38, Ca^{2+} = 110, Mg^{2+} = 33, K^+ = 1	B	[34]

Place	T	pH	M_T	S_T	SiO_2	CO_2	Main ionic components (mg/L)	D	Ref.
Monchi-que	32	9.6	335	—	14	—	F^- = 1, HCO_3^- = 105, Cl^- = 33, SO_4^{2-} = 46, Na^+ = 107, Ca^{2+} = 1, K^+ = 2	Su	[35]
Malhada Quente*	28	9.0	335	—	11	—	HCO_3^- = 181, Cl^- = 24, SO_4^{2-} = 16, Na^+ = 73, Ca^{2+} = 20, Mg^{2+} = 1	B	[35]
Alferce (Faro)*	27	8.6	607	—	3	—	HCO_3^- = 139, Cl^- = 101, SO_4^{2-} = 141, Na^+ = 184, Ca^{2+} = 3	Su	[35]
Santo Antó*	25	6.9	745	—	15	—	HCO_3^- = 459, Cl^- = 67, SO_4^{2-} = 11,Na^+ = 49, Ca^{2+} = 109, Mg2 + =24	B	[35]

T—temperature at emergence (°C), M_T—total mineralization (mg/L), S_T—total sulfuration (in I_2 0.01 N—mL/L), SiO_2—colloidal free silica (mg/L), CO_2—free carbon dioxide (mg/L), D—typical designation as to chemical composition: B—Bicarbonate, *C*—Chloride, *F*—Ferruginous, *N*—Nitrate, *S*—sulfurous, *Su*—Sulfate.
*This site is not currently active.

Table 4.
Main physical-chemical parameters of special groundwater in zone III, classified as NMW, in use in thermalism.

results are also associated in isolated publications [8, 10, 36–38] of some new NMW, which have recently been legalized and currently serve new medical spas in thermalism.

From the previous studies, the results representing the main occurrences of NMW in thermalism, or strong potential to become so, were organized into three groups, presented in **Tables 2–4**, according to the territory where they occur (Zones I, II, III, **Figure 1b**). The organization presented in those tables follow the format followed by Frederico Teixeira in 2022 [21], based on Portugal's systematization in terms of the National Health Service (SNS), since thermalism is framed in the activities of this entity, according to the following:

- Zone I—region under the influence of the ARS of the North;
- Zone II—region under the influence of the ARS of the Centre;
- Zone III—region under the influence of the ARS of Lisbon and Tagus Valley, the ARS of Alentejo, and the ARS of Algarve.

ARS corresponds to Regional Health Administration authorities.

To better understand the types of waters present, **Figures 3–7** are presented to show the variability of M_T, pH, and the types of dominant ions.

According to the M_T classification (**Table 1**): *weakly mineralized* waters are the most frequent, with M_T between 200 and 1000 mg/L; *hyposaline* waters (M_T < 200 mg/L) are very few, being only Luso, Monfortinho, Envendos, and Caldelas; *mesosaline* waters (1000 < M_T < 2000 mg/L) are very rare, with only two, Melgaço and Castelo de Vide; and the *hypersaline* waters (M_T > 2000 mg/L) have some significance, namely in Zone III, where Estoril is located, which is the most mineralized water of the current NMW, with M_T = 5743 mg/L.

Regarding the pH (**Figure 4**), most of the waters are *basic* or *alkaline*, that is, pH higher than 7, and it should be noted that the two waters with the highest value occur in Zone III, Monchique and Cabeço de Vide, with pH of 9.6 and 11.4, respectively. It should be noted that some *acidic* waters also occur, being the lowest values in Envendos, and Luso, with 4.7 and 5.4, respectively.

Regarding the main ions (**Figure 5**): (i) the Zone I waters are mostly of the *bicarbonate sodium* type (Group V, **Figure 5a**), and it should be noted that it includes the three most mineralized waters: Chaves, Pedras Salgadas, and Vidago, which is a situation well evidenced in **Figure 5b** (samples 8, 16 and 22); (ii) in Zone II (**Figure 5c**), the majority of waters are also of the *bicarbonate sodium* type, although in this region, a greater diversity is evident than in Zone I; it should be noted that in this zone (II), the two most mineralized waters are Curia and Monte Real, being of the *sulfate calcium* type (samples 9 and 16, **Figure 5d**); (iii) in Zone III, the situation is different from the previous ones, since there is only one *bicarbonate sodium* water (Malhada Quente), and the majority are *chloride sodium*-type waters (Group II, **Figure 5e**), with Estoril, Sta Marta (Ericeira), Cucos, and Vimeiro as the most mineralized in this region and even Estoril, the most mineralized, currently in use in the thermalism of mainland Portugal. Mention should be made particularly that the least mineralized NMW in Portugal, Envendos, is also of the *chloride sodium* type, but this is not possible to show in **Figure 5f** (sample 3) due to the scale of the diagram imposed in order to show the best relationship between most of the samples.

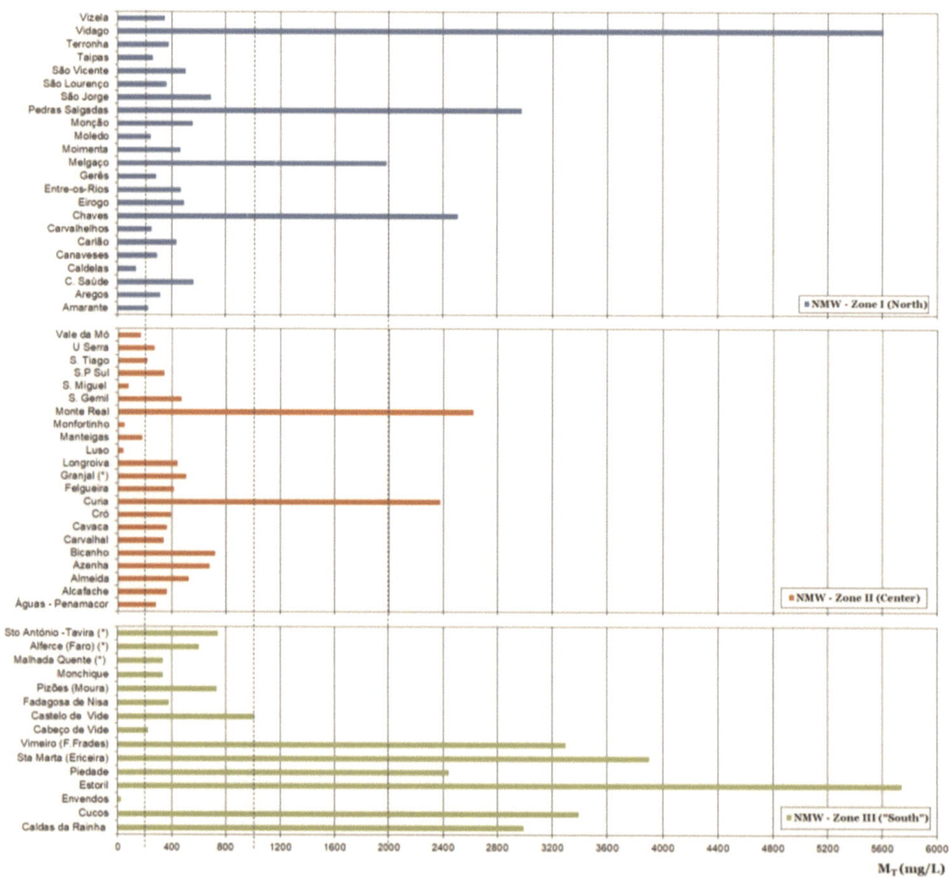

Figure 3.
Total mineralization (M_T) of the special groundwaters of mainland Portugal, classified as NMW, in thermalism activity. (*) *The water of this site, having the potential to be NMW, is not currently in thermalism.*

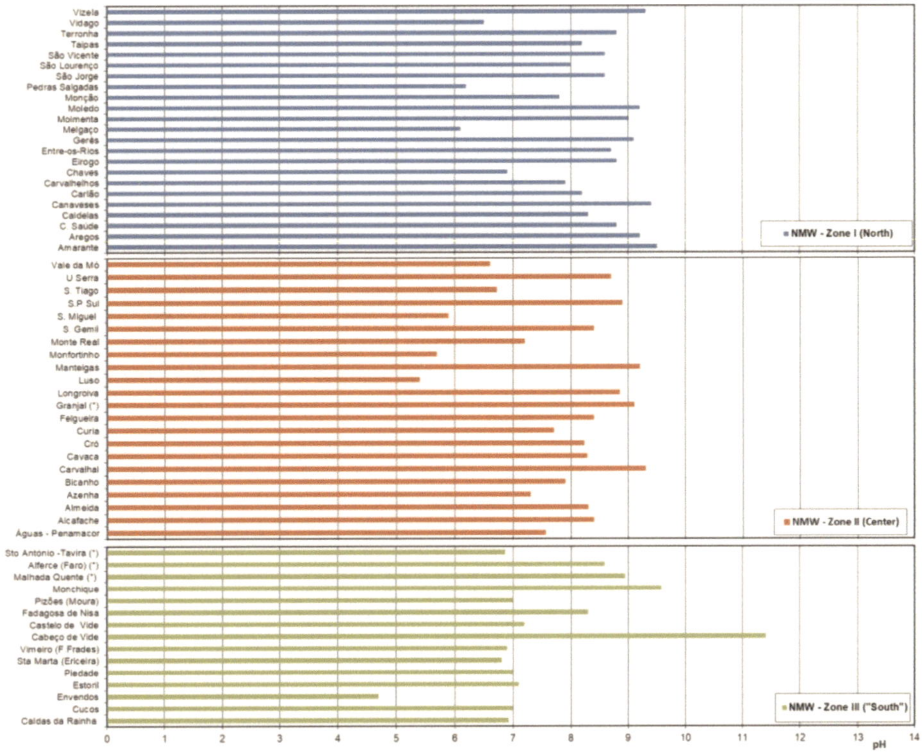

Figure 4.
pH of the special groundwaters of mainland Portugal, classified as NMW, in thermalism activity. (*) *The water of this site, having the potential to be NMW, is not currently in thermalism.*

Regarding the classification of the various Portuguese NMW and, in particular, the classification that each NMW is best known, presented in **Tables 2–4**, it is now of interest to highlight the same, which in general is a consequence of the most evident components due to their specific singularities. For example, waters that appear as *sulfurous* in most situations are *weakly mineralized*, *alkaline*, and *bicarbonate sodium*, but the singularities of these waters, because of the various associated forms of sulfur and represented by total sulfuration (S_T), give them a special character, namely the fact that they smell like rotten eggs and present a whitish cream on the surface near the emergences.

Thus, in relation to the totality of the NMW in mainland Portugal, the majority are of the *sulfurous* type, namely in the northern region (Group I), with the São Vicente NMW having the highest value, with S_T = 162 mg/L. A similar situation to the previous one is considered for fluoride, as all *sulfurous* waters are also *fluoridated*. Regarding Fe^{2+}, only two Portuguese NMW waters appear in that condition: São Tiago (Fe^{2+} = 3 mg/L) and Vale da Mó (Fe^{2+} = 9 mg/L), although only the latter is known as *ferruginous*.

About the free SiO_2, the situation in Portugal is amazing, since almost all waters have SiO_2 higher than 10 mg/L, namely almost all *sulfurous* waters, noteworthy being that the most silicate is Caldas da Saúde, with 94 mg/L. Still on silica, it is interesting to note that in the Portuguese NMW, the commonly known as *silicate* waters, are *hyposaline* waters, as in the cases of Luso, Monfortinho, and Envendos (**Tables 2** and **3**), with SiO_2

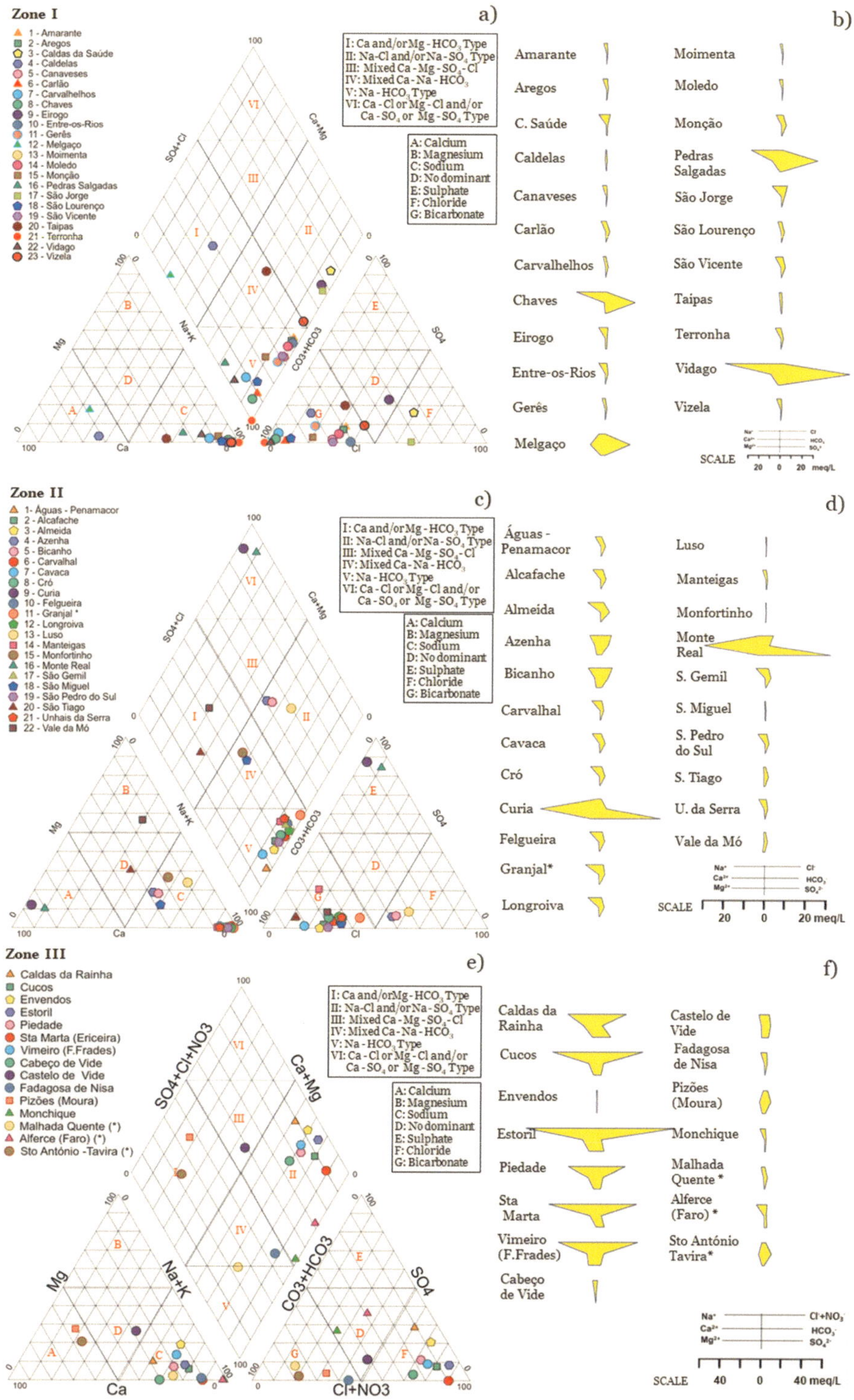

Figure 5.
Piper (a) and stiff (b) diagrams of special groundwater of zones I, II, and III, classified as NMW, in thermalism activity. (*) The water of this site, having the potential to be NMW, is not currently in thermalism.

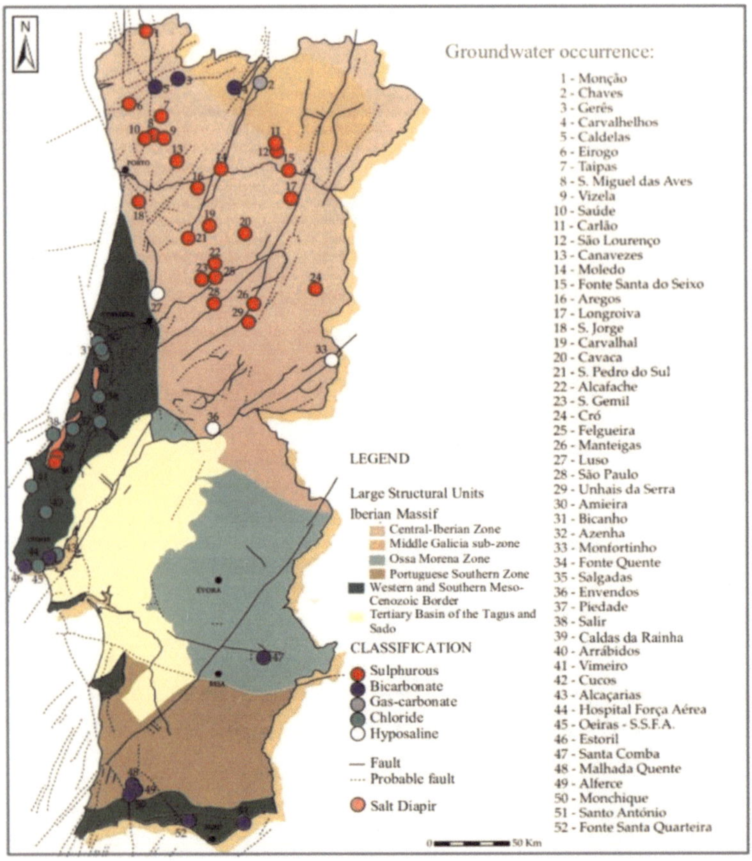

Figure 6.
Main Portuguese groundwater occurrences with geothermal potential of which most are NMW from ref. [39].

between 10 and 18 mg/L, but note that despite those values being low, in relative terms, they have a lot of meaning in relation to M_T, as these waters have M_T between 26 and 49 mg/L, only. Regarding the occurrence of free CO_2, in mainland Portugal, it is not a very common situation; however, it is a special singularity in the NMW of Chaves, Melgaço, Pedras Salgadas, and Vidago, all in the Northern Region (Group I), with CO_2 values between 1150 and 2550 mg/L; these NMWs are commonly called *gas-carbonates* (**Table 2**).

By the presented, it turns out that there is a considerable variety of types of groundwater. In global terms, the chemical type of groundwater is a consequence of the water-rock interactions, which initially were meteoric, evolve in depth on their path, until they resurface in natural springs or in abstractions strategically well implanted. In some cases, the NMWs, in addition to serving the thermalism, are also used in other applications, such as serving for drinking water, examples being the waters of Luso, Pedras Salgadas, Carvalhelhos, and Monchique [34].

A generic notion on the relationship between the type of groundwater and the regional geology can be observed in **Figure 6**, which presents on the geological map of Portugal, the location of the main Portuguese groundwaters, mostly in exploration as NMW. From that to **Figure 6**, it is emphasized that there are some groups of groundwaters to be highlighted: the *sulfurous* groundwater, which occur in the Central-

Iberian zone, whose rock masses are of granitoid type; the *chloride* groundwaters that occur in the Western Meso–Cenozoic border, whose rock masses are of sedimentary type, ranging from carbonate rocks to detrital rocks; and the *hyposaline* groundwaters that occur also in the Central-Iberian zone, with the particularity of being associated with rock masses of quartzite type of the Ordovician.

Considering the main goal of this chapter, to enhance the use of thalassotherapy from salty groundwater, indirectly coming from the sea, it is emphasized at this stage that some of the Portuguese NMW, from the previously mentioned, already have a large component of Cl and Na, typical of seawater. This happens as seen in particular in Estoril water (**Table 4** and **Figure 5e** and **f**), which is the most mineralized water currently used in thermalism, in Portugal. In any case, it is emphasized that there are immense chloride waters in the Western and Southern Meso-Cenozoic border (**Figure 6**). Also emphasized is the fact that in this region are associated geological formations of the Salt Diapir type and that in certain hydrogeological situations allow occurrences of natural, very salty groundwater; this is the case of Salgadas groundwater (point 35, **Figure 6**), which is currently not in thermalism use but corresponds to a concession, designated HM-65—Termas Salgadas da Batalha [40] and have an M_T of 32,500 mg/L, corresponding to a *hypersaline* water, of the Cl-Na type. This water has been used in the past in thermalism for rheumatic lesions, lymphatic disorders, ladies' diseases, horny skin lesions, consequences of traumatic lesions, and scrofula [41].

3.2 Physical: chemical aspects of salty groundwater in Portugal

This item serves to explore the physical–chemical aspects of salty groundwaters not currently licensed as NMW, namely the salty groundwaters of the Algarve (**Figure 7**) and the "Salgadas da Batalha" groundwater. The latter had already served in the past in thermalism activities; there are indications that it will soon be reactivated, with new studies not only around chemistry but also on its therapeutic and other, as guided by the recent study on the hydrogenome [40].

Thus, **Table 5** presents the main physical–chemical parameters of those waters, available in the literature [30, 41]. It is mentioned that since M_T is not available, and there is a need to compare with the situation of NMW, that parameter is evaluated from [27]:

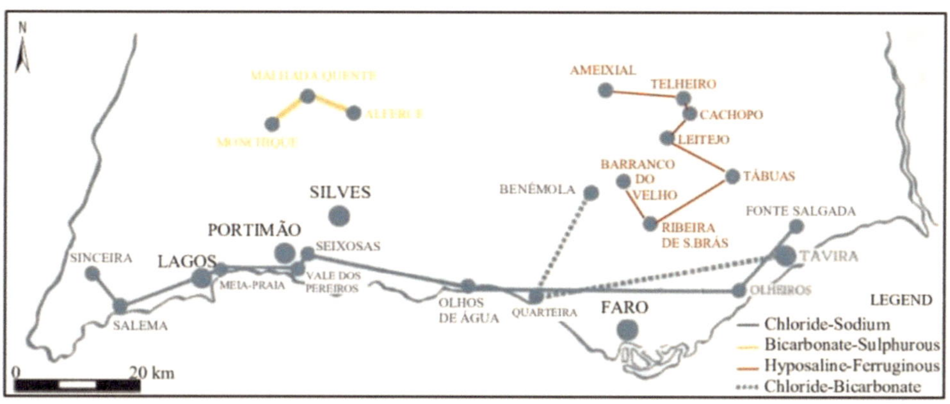

Figure 7.
Singular groundwater occurrences in the Algarve [30].

Place	Type	pH	R_D	CO_2	SiO_2	Main ionic components (mg/L)	Ref.
Sinceira (Vila do Bispo)	Fontain	5.3	289	119	84	Cl^- = 21, HCO_3^- = 12, SO_4^{2-} = 8, NO_3^- = 38, Na^+ = 66, Ca^{2+} = 10, Mg^{2+} = 15,	[30]
Salema (Budens)	Hole	6.9	3436	77	12	Cl^- = 1702, HCO_3^- = 403, SO_4^{2-} = 296, NO_3^- = 40, Na^+ = 970, Ca^{2+} = 158, Mg^{2+} = 130	[30]
Meia-Praia (Lagos)	Hole	7.2	936	29	10	Cl^- = 355, HCO_3^- = 268, SO_4^{2-} = 91, NO_3^- = 20, Na^+ = 188, Ca^{2+} = 107, Mg^{2+} = 25	[30]
Seixosas (Ferragudo-Lagoa)	Hole	7.0	3481	68	24	Cl^- = 1683, HCO_3^- = 383, SO_4^{2-} = 240, NO_3^- = 2, Na^+ = 916, Ca^{2+} = 197, Mg^{2+} = 110	[30]
V. dos Pereiros (Ferragudo-Lagoa)	Hole	7.0	2202	68	37	Cl^- = 994, HCO_3^- = 395, SO_4^{2-} = 180, NO_3^- = 7, Na^+ = 575, Ca^{2+} = 146, Mg^{2+} = 73	[30]
Olhos de Água (Albufeira)	Spring	7.0	1373	56	5	Cl^- = 557, HCO_3^- = 356, SO_4^{2-} = 95, NO_3^- = 4, Na^+ = 301, Ca^{2+} = 120, Mg^{2+} = 55	[30]
Olheiro (Fuseta-Olhão)	Spring	6.8	5577	89	24	Cl^- = 2911, HCO_3^- = 400, SO_4^{2-} = 409, NO_3^- = 1, Na^+ = 1587,Ca^{2+} = 234,Mg^{2+} = 199	[30]
Fonte Salgada (Tavira)	Spring	7.3	12,367	24	9	Cl^- = 6135, HCO_3^- = 307, SO_4^{2-} = 1417, Na^+ = 4233, Ca^{2+} = 44, Mg^{2+} = 124, K^+ = 40	[30]
Termas Salgadas da Batalha	Well	7.3	32,000	—	14	Cl^- = 17,333, HCO_3^- = 297, SO_4^{2-} = 1783 Na^+ = 11,083, Ca^{2+} = 424, Mg^{2+} = 58, K^+ = 14	[41]

R_D—Dry Residue at 180°C (mg/L), CO_2—free carbon dioxide (mg/L), SiO_2—colloidal free silica (mg/L).

Table 5.
Results of analyses of chloride sodium groundwater of the Algarve region and the Termas Salgadas da Batalha, Portugal.

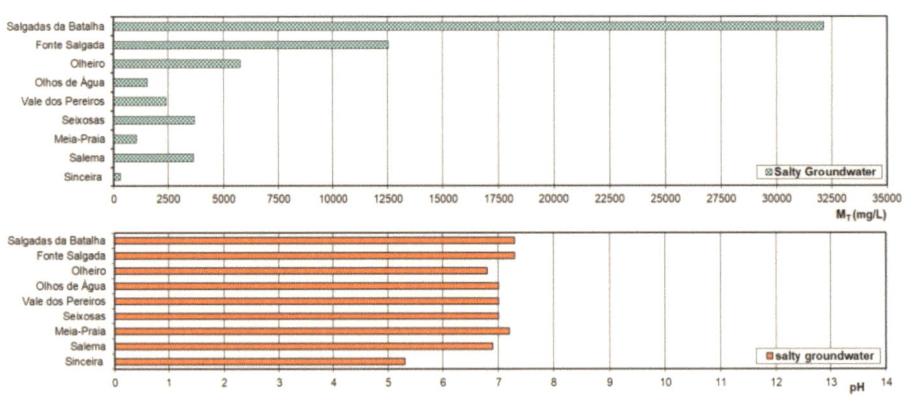

Figure 8.
Total mineralization (M_T) and pH of special salty groundwaters in mainland Portugal.

$$M_T \cong R_D + ½\, HCO_3^- \quad (1)$$

R_D being the Dry Residue, obtained at 180°C.

At the similar of the graphics presented for the NMW, **Figure 8** shows the variability of M_T and pH in the various salty groundwaters, and **Figure 9** shows the Piper

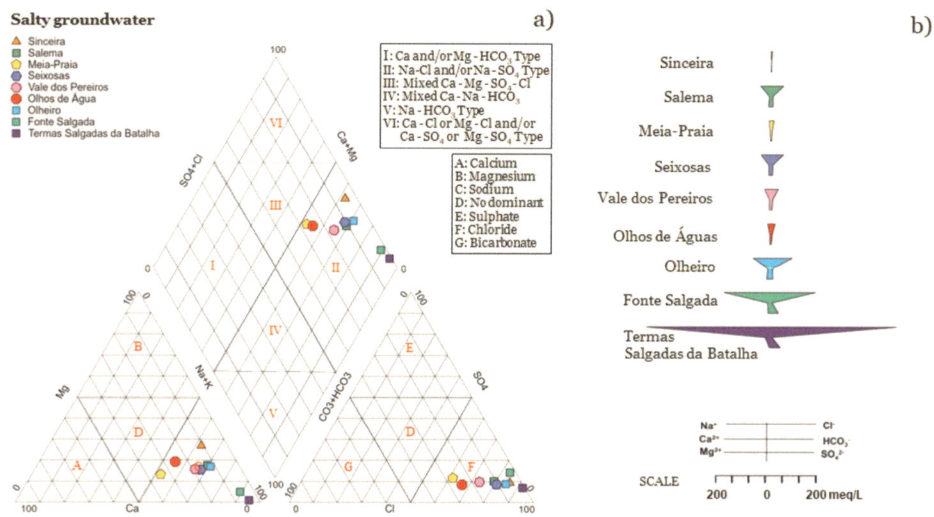

Figure 9.
Piper and stiff diagrams of the special salty groundwaters of mainland Portugal.

and Stiff diagrams of the same waters. Regarding M_T, most waters are *hypersaline* ($M_T > 2000$ mg/L), only two (Meia Praia and Olhos de Água) are *mesosaline* (M_T between 1000 and 2000 mg/L), and only one (Sinceira) is *weakly mineralized* (M_T between 200 and 1000 mg/L). With regard to pH, with the exception of Sinceira, all have a pH of around 7, that is, neutral. Regarding the ionic composition (**Figure 9**), all these waters fit into the situation where Cl and Na ions are in the majority, integrating exclusively sector II of the Piper diagram, and are therefore *chloride sodium*-type waters.

3.3 Physical-chemical aspects of seawater

The temperature of seawater varies from −4°C in the Arctic to 30°C in tropical regions, and the density of seawater varies from 1028 mg/L to 1032 mg/L [15]. In

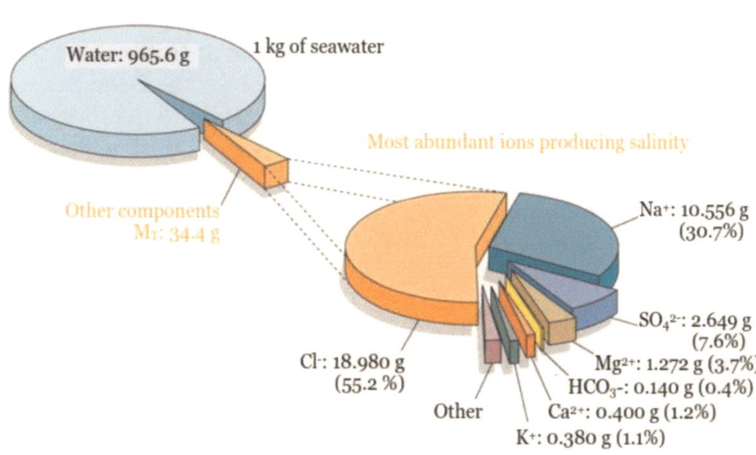

Figure 10.
Major chemical constituents of seawater [43].

Sea or Ocean	Salinity / Total mineralization—M_T (mg/L)	Ref.
Baltic Sea	7000	[44]
Black Sea	13,000	[44]
Adriatic Sea	25,000	[44]
Pacific Ocean	33,000	[44]
Indian Ocean	33,800	[44]
Atlantic Ocean	36,000	[44]
Mediterranean Sea	39,400	[44]
Arabian Gulf	43,000	[44]
Red Sea	43,000	[44]
Dead Sea	300,000–350,000	[21]

Table 6.
Order of magnitude of total mineralization (M_T) for saline waters from different oceans and seas.

chemical terms, the major constituents in seawater are Cl^-, Na^+, Mg^{2+}, SO_4^{2-}, Ca^{2+}, K^+, Br^-, and Sr^{2+}, and the minor constituents are several, namely: PO_4^{3-}, NO_3^-, and SiO_2 [42]. Trace elements, in the order of μg/L, exist in great variety, and it can be said that seawater has practically all the elements of nature. **Figure 10** shows the participation, in relative weight terms, of the six most abundant chemical elements or mineral salts that, in ionic form, are present in 1 kg of seawater [43].

The physical–chemical composition may vary depending on the ocean or sea. The most common situation in oceans in terms of total salinity or total mineralization (M_T) is around 33 to 36 g/L; however, as can be seen in **Table 6**, considering seas and other saltwater masses, M_T can have values from 7 g/L in the Baltic Sea to 350 g/L in the Dead Sea. Seawater, especially ocean water, also contains a large quantity of gases, about 20 to 30 cm^3/L, mainly oxygen, hydrogen, and carbon dioxide [21]. The concentrations of chemical species in seawater do not vary much, either horizontally or vertically, except in river-mouth regions where a greater quantity of salts brought by rivers occurs [42].

To have a notion of its detailed chemistry, and to compare it with NMW and other Portuguese salty groundwaters, the main physical–chemical parameters of some waters from various oceans and seas are presented in **Table 7**, based on results available in the literature. **Figure 11** shows the Piper and Stiff diagrams of the same waters. From these diagrams, it can be said that, except for the Dead Sea water, all the others fall into Group II as *chloride sodium* waters in relation to the Piper diagram. The waters of the Dead Sea are in Group VI, as chloride magnesian. The fact that they belong to a different chemical group is also evidenced in the Stiff diagram (**Figure 11b**), and the proximity of the Mediterranean waters with those of the Atlantic Ocean and even with the water pattern of the Indian Ocean is also emphasized.

3.4 Relationships of physical-chemical parameters between NMW, special salty groundwater, and seawater

To investigate possible relationships between the various physical–chemical parameters, the relationship between M_T and conductivity (parameters taken from the references mentioned in **Tables 2–4**) was first studied for all the NMW. The relationship obtained is shown in **Figure 12a**, with the following expression:

Parameter		Atlantic Ocean [43]	Mediterranean Sea [43]	Dead Sea [43]	Dead Sea [45]	Mediterranean [46]	Indian Ocean [47]
pH		—	—	8.05		7.5	7.57
M_T* (mg/L)		≈ 35,554	≈ 37,876	≈ 372,874	≈ 322,130	≈ 36,509	24,602
Anions (mg/L)	Cl^-	18,981	20,980	239,287	212,400	19,678	12,623
	HCO_3^-	140	155	8211	220	137	2814
	SO_4^{2-}	2470	2720	536	470	2782	1177
	NO_2^-	—	—	—		—	0.53
	NO_3^-	—	—	16		10	3.0
	CO_3^{2-}	—	—	332		—	—
	F^-	1.3	—			1.5	0.06
	Br-	65	72		5120	6.4	
	$H_3BO_3^-$	24	27	—		—	—
	$H_2PO_4^-$	—	—	—		—	0.07
Cations (mg/L)	Na^+	10,540	11,650	42,090	39,150	11,496	5500
	Mg^{2+}	1270	1410	47,142	40,650	1396	624
	Ca^{2+}	400	442	**	16,860	540	1560
	K^+	380	420	18,400	7260	462	300
	Sr^{2+}	13	—	—		—	—
	Fe^{2+}	—	—	—		—	0.01
	Li+	—	—	—		0.09	—

**the values presented as approximate (≈) resulted from the sum of the various ionic components presented in this table. **for the purpose of M_T calculation and construction of Piper and Stiff diagrams, it was considered for Ca^{2+}, the same value presented in Ref. [45].*

Table 7.
Results of physical–chemical analyses of waters from different oceans and seas.

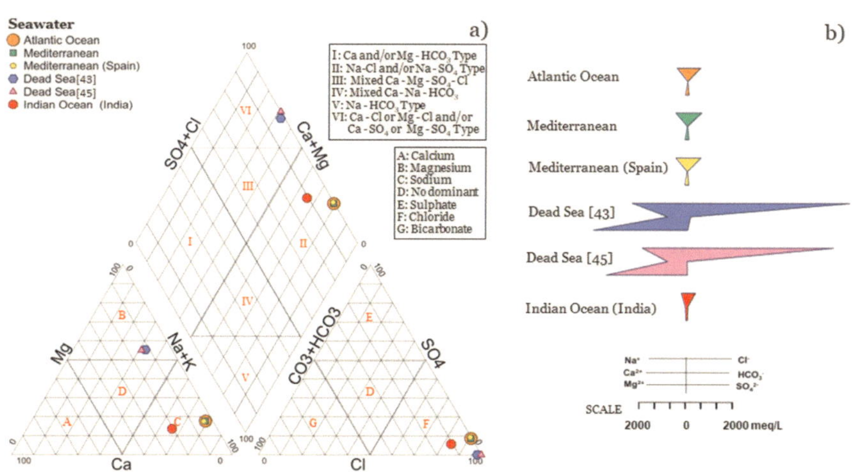

Figure 11.
*Piper (a) and stiff (b) diagrams of waters from different oceans and seas (from **Table 7**).*

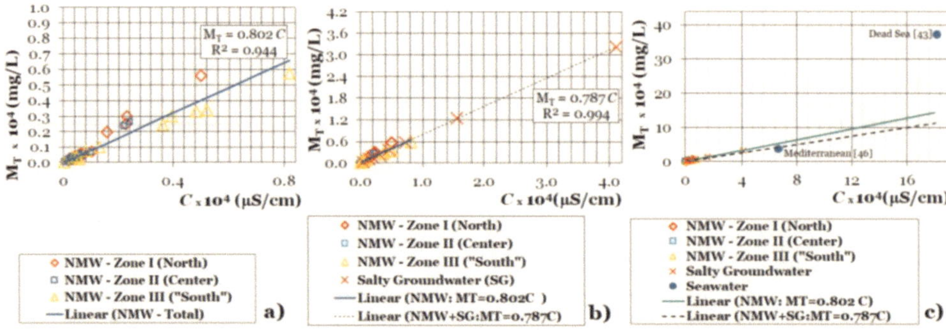

Figure 12.
Relationships between conductivity and total mineralization in the various waters under study: (a) natural mineral waters (NMW), (b) NMW and salty groundwater, and (c) NMW, salty groundwater, and seawater.

$$M_T = 0.802\,C \tag{2}$$

where C is conductivity in μS/cm, and M_T is total mineralization in mg/L.

Above the trend line (**Figure 12a**) are clearly located the waters of Zone I, with the *gas-carbonate* classification (**Table 2**), namely, Vidago, Pedras Salgadas, Chaves, and Melgaço; in ionic terms, the first three are of the *bicarbonate sodium* type, and the last one is of the *bicarbonate calcium* type; all four of these waters are singular in the Stiff diagrams (**Figure 5b**). The waters classified as *sulfate* in Zone II (**Table 3**—Curia and Monte Real) are still above the global trend; these waters are, in ionic terms, of the *sulfate calcium* type (**Figure 5c**), and it should be noted that, according to the Stiff diagram, they are unique in relation to the others (**Figure 5d**). It should also be noted that below that trend (**Figure 12a**) are the *chloride* waters, of the Zone III (**Table 4**), ionically all *chloride-sodium* (Group II, Piper diagram, **Figure 5e**), which are also unique in the Stiff diagram (**Figure 5f**).

In order to have a global notion of the position in that type of graph, of salty groundwater (**Table 5**), the M_T evaluated from Eq. (1) was used, and Eq. (2) was admitted as valid for this type of groundwater of the reference [30] in **Table 5**; there is the particularity of the experimental result of the water of reference [41], which is C = 41,100 μS/cm. It results like this **Figure 12b**, with a trend very close to that obtained for the NMW and which is as follows:

$$M_T = 0.787\,C \tag{3}$$

with C in μS/cm and M_T in mg/L.

Also, to compare with the seawaters, although only direct results of M_T and C, from two waters (Mediterranean Sea and Dead Sea), are available, **Figure 12c** is presented. This highlights the exceptionality of the point corresponding to the Dead Sea. Anyway, it is interesting to verify that the Mediterranean Sea water position is very close to the trend of the "NMW + Salty groundwater," being remarkable, that according to the physical–chemical composition of the other seas waters (**Tables 6** and 7), excluding therefore the Dead Sea water, very special indeed, the general trend of M_T versus C, is close to Eq. (3).

In order to analyze the situation between the various types of waters in ionic terms, the pH situation with HCO_3^- and Na^+ is shown in **Figure 13** as an example. The graphs in that figure show that there is no pH trend with those ions. A similar situation occurs with the other major ions (Cl^-, SO_4^{2-}, Ca^{2+}).

DOI: http://dx.doi.org/10.5772/intechopen.109763

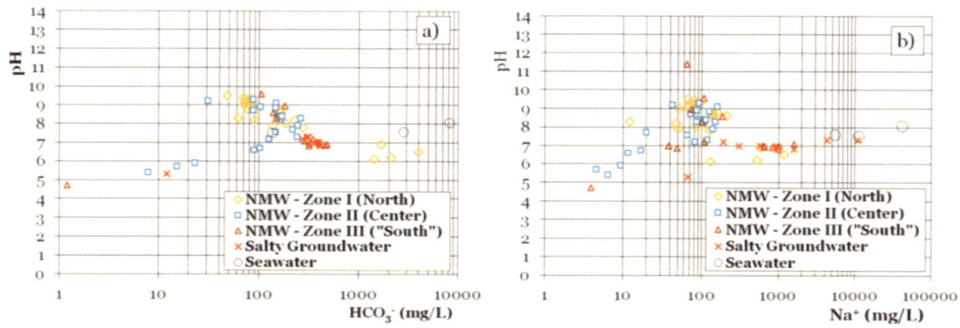

Figure 13.
Relationships between pH and ions are often very important in the various waters under study.

The relationships between M_T and the most important ions in the various waters under study were investigated, resulting in some interesting relationships on a *bi-log scale*, as shown in the graphs in **Figure 14**. There is a clear contribution of Na^+, Cl^-, Ca^{2+}, and Cl^- in the increase of M_T in all the studied waters (NMW, salty groundwater, and seawater). The case of HCO_3^- clearly contributes to the increase of M_T for most of the NMW, but this does not happen for the salty groundwater and also for the seawater.

In relation to the main ions, **Figure 15** shows some graphs of those considered most important. The increasing of Na^+ with increasing Cl^- is a very evident situation for most of all waters under study; the situation of increasing Na^+ and Ca^{2+} with increasing HCO_3^- in global terms is also true, but there are many exceptions, namely for the case of Na^+ in salty groundwater and seawater and even more exceptions for the case of Ca^{2+}. In the case of the relationship between HCO_3^- and Cl^-, the situation is very

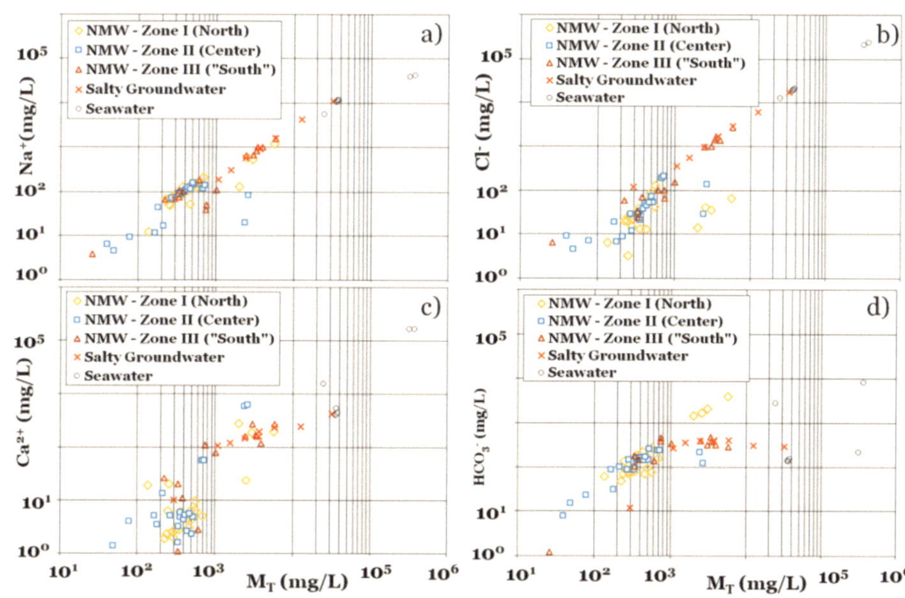

Figure 14.
Relationships between M_T and ions are often very important in the various waters under study.

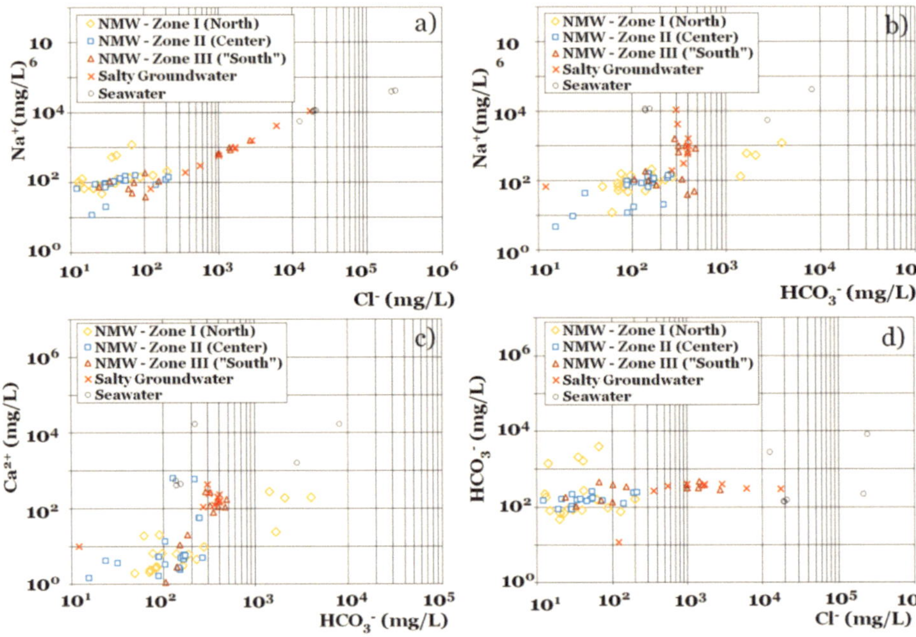

Figure 15.
Relationships between various ions are often very important in the various waters under study.

clear that these ions are not related to each other (**Figure 15d**); independently of the Cl^- value, HCO_3^- in most situations occurs with values between 100 and 1000 mg/L.

4. Conclusions and final notes

In Portugal, there is a wide variety of natural mineral waters (NMWs) in chemical terms (**Figure 6**), including some relatively salty waters (rich in Cl and Na), with Estoril water being the most mineralized water in Portugal (**Figure 3**), currently in operation for thermalism, with a total mineralization (M_T) of 5743 mg/L (**Table 4**).

It is emphasized that NMW, according to Portuguese legislation, is currently a groundwater, which, among other factors, presents a high physical–chemical stability over time, regardless of the M_T value.

There are some very salty groundwaters (rich in Cl^- and Na^+) in Portugal, namely in the Algarve region (**Figure 7**) and also in the Western Meso-Cenozoic Border region (**Figure 6**), associated in particular with geological formations of the salt diapir type, as is the case of the Salgadas da Batalha water, with an M_T of about 32148.5 mg/L, and it should be noted that this has already been used in thermalism.

On saline waters of various oceans and seas, they are mostly composed of Cl^- and Na^+ ions (**Figure 10**), with M_T varying between orders of magnitude from 7000 mg/L in the Baltic Sea to 43,000 mg/L in the Arabian Gulf and Red Sea (**Table 6**).

In the Atlantic Ocean, which coexists with western mainland Portugal, the M_T takes on values of the order of 36,000 mg/L, and in the Mediterranean Sea, which coexists with Portugal, to the south, in the Algarve, the M_T takes on the order of magnitude of 39,400 mg/L. The Dead Sea water reaches very high values of M_T, from 300,000 to 350,000 mg/L (**Tables 6** and 7), and the two major ions are Cl^- and Mg^{2+} (**Table** 7, **Figure 11**).

The sea water, in Portugal, for use in thalassotherapy, is captured directly from the Atlantic Ocean or from the Mediterranean Sea. Some thalassotherapy spas already have the supervision of a doctor with a specialization in Hydrological Medicine and therefore, in some situations, have activities for users very similar to those of thermalism.

Thermalism, which has very specific and very clear legislation, uses the NMW, highlighting that these are special groundwaters, namely for presenting high physical–chemical stability over time, and Portuguese law requires that their exploitation, from the geological environment, is the responsibility of a professional in the field of hydrogeology.

Thalassotherapy in Portugal has no specific legislation. Its activity is guided in accordance with general laws applied to the sector regulating the exploration of municipal swimming pools, private swimming pools, spas in hotels, private spas, and tourist complexes, among others.

Thus, Portugal, having already in use some NMW of the chloride sodium type, if it has, in the future, salty groundwaters, captured from the aquifer system in contact with the sea, and if, after all the studies, legally imposed for the NMW, come to demonstrate adequate quality, it is admitted that without any doubt in the future the Portuguese thalassotherapy network can be integrated in the network and rules of thermalism in Portugal.

The thermalism sector will grow, and the thalassotherapy sector, with great probability, will have salty waters of greater stability in quality. Thalassotherapy water would no longer be abstracted in the liquid wedge of the sea, where it normally has great oscillations, namely of vestigial chemical species and remains of hydrocarbons and oils, among others, but be abstracted as classic groundwater. In this way, the water to be exploited, before entering the well (groundwater abstraction), passes through the geological environment, which acts with its natural purifying power, to eliminate potential harmful elements that destabilize its quality.

The construction of groundwater abstractions must follow the good rules of hydrogeology, with a particularity of having the notion of not causing the development of the saline wedge to the interior of the territory, as shown in many works in this scientific field [48, 49].

In zones of free (unconfined) aquifers, in loose sands, an abstraction such as the one shown in **Figure 2** can be built, which minimizes this problem. However, it is acceptable to build boreholes or classic wells with the aid of roto percussion, percussion, or other drilling machines but with due care, both in drilling and in pumping operations during flow tests, especially in the prospecting phase.

The groundwater abstractions of the thalassotherapies, after their official classification as NMW, would have the compulsory implementation of a protection perimeter, with three zones surrounding the abstraction (Immediate, Intermediate, and Extended Zone) to be subject to continuous "supervision" over the occupation of the territory and anthropic actions in it, besides having the obligation to include a hydrogeologist, the Technical Director, as responsible for the exploration of that resource. Such situations will be guarantees for the problem of the control of the geometric stability of the wedge-saline, namely in the eventual prohibition of new abstractions of groundwater without rules and control for third parties, whether in tourism, agriculture, or other activities.

A major advantage when evolving towards this type of solution is the possibility of obtaining stable exploration flows over time, from salty groundwater abstractions, close to the sea. It should be noted that the stability of exploitation flows of the NMW is starting to be in danger due to climate change, since it is common sense that if normal hydrological cycles do not occur, that is, with the current trends of less precipitation and more evaporation, the sustainable exploitation of aquifer systems will be put in danger, leading necessarily to a decrease in exploitation flows and causing consequent potential changes in the quality of the resource.

Thalassotherapies supplied by salty groundwater, coming indirectly from the sea, can be an excellent way for the future of thermalism, which already has a millenary tradition, and it is necessary to guarantee its continuity and consolidation in this new phase of humanity that faces the fact of climate change in many regions of the world.

Acknowledgements

FCT—Foundation for Science and Technology supported this work with Portuguese funds within the GeoBioTec Centre (Project UIDB/04035/2020).

Conflict of interest

The data sets generated during and/or analyzed during the current study are available from the corresponding author on request after the publication of work.

Author details

Luís Manuel Ferreira-Gomes[1,2]*, Luís José Andrade Pais[1,2] and Pedro Jorge Coelho Ferreira[3,4]

1 Beira Interior University (UBI), Covilhã, Portugal

2 Centre GeoBioTec (UA), Aveiro University, Aveiro, Portugal

3 Municipality of Meda, Meda, Portugal

4 Department of Civil Engineering and Architecture, Beira Interior University (UBI), Covilhã, Portugal

*Address all correspondence to: lmfg@ubi.pt

References

[1] Direção-Geral do Território. Carta Administrativa Oficial de Portugal (CAOP), shapefile, DGT, Lisboa. 2017

[2] DIVA-GIS. Global country boundaries (Arcgis by ESRI). Global country boundaries from DIVA-GIS, shapefile. 2019. Available from: https://www.arcgis.com/home/item.html?id=2ca75003ef9d477fb22db19832c9554f

[3] Portuguese Decree. Aprova Lei das Termas. Dec.-Lei n°142/2004, D.R., I Série—A, N°136, Lisboa, 2004, pp. 3632-3640

[4] ATP, Associação de Termas de Portugal. 2022. Available from: http://www.termasdeportugal.pt

[5] Portuguese Decree. Approves the natural mineral water regulation. Decreto-Lei n°86/90, D.R. 63, series I-A, Lisboa, 1990, pp. 1254-1264

[6] Portuguese Decree. Approves the bases for the legal regime of the revelation and exploitation of geological resources. Lei n°54/2015, D.R., 119, I Série, Lisboa, 2015, pp. 4296-4308

[7] Ferreira-Gomes LM, Carvalho, PEM, Andrade Pais L. Estudos hidrogeológicos para a classificação de uma nova água mineral em Portugal—As Termas de São Miguel, Fornos de Algodres. 14.° Cong. da Água. 7–9 de março, APRH, Évora, 2018, p. 15, ISBN 978-989-8509-21-6

[8] Ferreira-Gomes LM, Maia de Carvalho PE, Andrade Pais LJ. Hydrogeochemical studies of a groundwater with a view to its classification as mineral water for a new medical spa in Portugal. IOP Conference Series: Earth and Environmental Science. 2019;**221**:012030. DOI: 10.1088/1755-1315/221/1/012030

[9] Ferreira-Gomes LM. Main geoenvironmental studies to launch new medical spa: The case of the São Tiago medical spa. Modern Environmental Science and Engineering, Academic Star Pub. Comp. 2020;**6**(7):723-736. DOI: 10.15341/mese(2333-2581)/07.06.2020/001

[10] Ferreira-Gomes LM. Quality of an underground hydro resource in order to classify it as natural mineral water for thermalism: The case of the new resource "Termas de São Tiago". IOP Conference Series: Earth and Environmental Science. 2020;**609**:012069. DOI: 10.1088/1755-1315/609/1/012069

[11] Lourenço C, Ribeiro L. Classificação das águas minerais naturais e de nascente de Portugal segundo as suas características físico-químicas. In: 7° Cong. da Água. 8-12 March 2004; Lisboa. APRH. ISBN 972-97480-6-3. 2004, p. 11

[12] EN, Commission Directive 2003/40/EC of 16 May 2003. Official Journal of the European Union. 2003. No L 126: 34-39. Available from: https://eur-lex.europa.eu/legal-content/PT/TXT/?uri=CELEX%3A32003L0040

[13] Valenzuela MA, Bacaicoa JSM. Curas balneares y Climáticas—Talasoterapia y Helioterapia. 1ª ed., Madrid: Editorial Complutense; 1994. p. 688

[14] Gomes CSF, Silva JBP, Fernandes JV, Fernandes FMV. Thalassotherapy in Porto Santo Island of the Madeira archipelago: Facts and prospects. Boletín de la Sociedad Española de Hidrología Médica. 2019;**34**(1):9-33. DOI: 10.23853/bsehm.2019.0953

[15] Gomes CSF. Talassoterapia, Rev. Ciência Elem. Univ. do Porto. 2021; **9**(04):067. DOI: 10.24927/rce2021.067

[16] Thalasso Costa da Caparica, Health—Wellness-Holistic. [Internet]. 2021. Available from: https://www.thalassocaparica.com/en/home

[17] Thalasso Nazaré .The healing that comes from the sea. [Internet]. 2021 Available from: https://thalassonazare.com/en/, 2021

[18] Vilalara Thalassa Resort; New Treatments. [Internet]. 2021. Available from: https://www.vilalararesort.com/en/

[19] Real Spa Therapy. Thalassotherapy. 2021. Available from: https://www.realspatherapy.com/en/thalassotherapy

[20] Gomes CSF, Fernandes JV, Fernandes FV, Silva JBP. Salt mineral water and thalassotherapy. In: Gomes C, Rautureau M, editors. Minerals Latu Sensu and Human Health; eBook; Cham: Switzerland: Springer; 2021. pp. 631-656. DOI: 10.1007/978-3-030-65706-2_16

[21] Teixeira F. Hidrologia Médica—Princípios Gerais. Lidel—Edições Técnicas, lda. Lisboa; 2022. p. 341

[22] Aldeia PC. Complexo Talasso-Termal no Cabo Espichel. Covilhã: Univ. of Beira Interior. p. 113. Available from: https://1library.org/document/z3127pmy-complexo-talasso-termal-do-cabo-espichel.html

[23] Aldeia PC, Ferreira-Gomes LM, Diniz FML.Contrıbuıções para um Complexo Talasso-Termal no Cabo Espıchel—Sesimbra". In: 8° Con. Luso-Moçambicano de Engenharia. 4-8 Sept. 2017; Maputo/Moçambique. 2017. pp.1095-1104. Available from: https://paginas.fe.up.pt/clme/2017/Proceedings/data/papers/6979.pdf

[24] Ferreira-Gomes LM, Aldeia PC, Teixeira F, Pais LJA. An impulse for the thalassotherapy activity in Portugal to be framed in activities of the medical spa type—A proposal for Cabo Espichel (Portugal). p. 9. In: 7th WMCAUS 2022, Praga, Czech Republic, 2022 (on press)

[25] Directive 80/778/EEC. Council Directive of 15 July 1980 Relating to the quality of water intended for human consumption. Official journal of the European Communities, 1980, No L 229: 11-29

[26] Rice EW, Baird RB, Eaton AD. (Eds.). Standard Methods for the Examination of Water and Wastewater, 23 Ed.; American Public Health Association: Washington, DC, USA, 2017; ISBN 978-16-2576-240-5

[27] Custodio E, Llamas MR. Hidrología Subterránea. 2nd ed. Barcelona: Omega; 1983

[28] Logiciels. Laboratoire d'Hydrogéologie d'Avignon. Logiciel d'hydrochimie—Water software quality hydrochemistry diagrams. Diagrammes. ZIP. 2022;**6**(5). Available from: http://www.lha.univ-avignon.fr/LHA-Logiciels.htm

[29] Henriques FF. Aquilégio Medicinal. Lisboa: Lisboa Occidental; 1726. p. 348. Available from: https://purl.pt/22614/1/index.html#/5/html

[30] Almeida A, Almeida JD. Inventário Hidrológico de Portugal—Algarve. Vol. 1. Lisboa: Instituto de Hidrologia de Portugal; 1966. p. 135

[31] Almeida A, Almeida JD. Inventário Hidrológico de Portugal—Trás-Os-Montes e Alto Douro. Vol. 2. Lisboa: Instituto de Hidrologia de Portugal; 1970. p. 640

[32] Almeida A, Almeida JD. Inventário Hidrológico de Portugal—Beira Alta.

Vol. 3. Lisboa: Instituto de Hidrologia de Portugal; 1975. p. 525

[33] Almeida A, Almeida JD. Inventário Hidrológico de Portugal—Minho. Vol. 4. Lisboa: Instituto de Hidrologia de Portugal; 1988. p. 505

[34] DGGM. Termas e Águas Engarrafadas em Portugal. Lisboa: Direção-Geral de Geologia e Minas; 1992

[35] IGM. Catálogo de Recursos Geotérmicos. Lisboa: Instituto Geológico e Mineiro, CD-Room; 1999

[36] Ferreira-Gomes LM, Antunes IMHR, Albuquerque MTD, Santos Silva AJ. New thermal mineral water from Águas (Penamacor, Central Portugal): Hydrogeochemistry and therapeutic indications. IOP Conference Series: Earth and Environmental Science. 2019;**221**: 012025. DOI: 10.1088/1755-1315/221/1/012025

[37] Dias Morgado SL. O termalismo e a saúde oral—um estudo observacional nas Termas de Amarante [thesis]. Viseu: Univ. Católica; 2022. p. 110. Available from: https://repositorio.ucp.pt/bitstream/10400.14/38776/1/203052510.pdf

[38] Ferreira-Gomes LM, Guedes JF, Costa TCG, Coelho Ferreira PJ, Trota APN. Geothermal potential of portuguese granitic rock masses: Lessons learned from deep boreholes. Environment and Earth Science. 2015; **73**(6):2963-2979. DOI: 10.1007/s12665-014-3605-y

[39] IGM. Recursos Geotérmicos em Portugal Continental, Baixa Entalpia. Ed. IGM, Lisboa, 1998. p. 23. Available from: https://www.dgeg.gov.pt/media/4dqalyio/i007552.pdf#Publica%C3%A7%C3%A3o_I007552_RecursosGeot%C3%A9rmicosEmPortugalContinentalBaixaEntalpia_IGM_1998

[40] DGEG. Hidrogenoma, Uma Nova Visão Estratégica para as Águas Minerais Naturais. Lisboa: Direção-Geral de Geologia e Energia; 2019. p. 197. Available from: https://www.dgeg.gov.pt/media/qjlpphdr/publica%C3%A7%C3%A3o_dsrhg_livro_hidrogenoma_2019_final_site.pdf

[41] Medeiros A, Teixeira E. Conceptualização do circuito hidromineral das Termas Salgadas da Batalha. In: 7° Sem. sobre Águas Subterrâneas, 5-6 March 2009. Lisboa, APRH–Associação Portuguesa de Recursos Hídricos, LNEC; 2009. p. 39

[42] Mason B. Principles of Geochemistry. N.York: John Wiley & Sons; 1962. p. 329

[43] Morer C. Talasoterapia. Bol. Sociedad Española de Hidrologia Médica. 2016;**31**(2):119-146. DOI: 10.23853/bsehm.2017.0209

[44] Desjardins R. Le traitement des eaux. Deuxième Édition. Québec: Presses Inter. Polytechnique; 1997. p. 304

[45] Bawab AA, Bozeya A, Abu-Mallouh S, Irmaileh BA, Daqour I, Abu-Zurayk RA. The dead sea mud and salt: A review of its characterization, contaminants, and beneficial effects. IOP Conference Series: Materials Science and Engineering. 2018;**305**:012003. DOI: 10.1088/1757-899X/305/1/012003

[46] Morer C, Michan-Doña A, Alvarez-Badillo A, Zuluaga P, Maraver F. Evaluation of the feasibility of a two-week course of aquatic therapy and thalassotherapy in a mild post-stroke population. International Journal of Environmental Research and Public Health. 2020;**17**:8163. DOI: 10.3390/ijerph17218163

[47] Eswari A, Amala B, Poonguzhali TV. Studies on the physico chemical analyses of two different water bodies. International Journal of Science, Environment and Technology. 2015; **4**(5):1377-1383

[48] Carrera J, Pool M, Abarca E, Hidalgo JJ, Slooten LJ, Vázquez-Suñe E, et al. Principios y conceptos básicos sobre intrusión marina y gestión de acuíferos costeros. In: Cuatro décadas investigación y formación aguas subterráneas. Libro Homenaje Prof. Emilio Custodio. Zaragoza 2011. AIH-GE; 2011. pp. 214-222

[49] Zachopoulos K, Kokkos N, Sylaios G. Salt wedge intrusion modeling along the lower reaches of a Mediterranean river. Regional studies in marine. Science. 2020;**39**:101467. DOI: 10.1016/j.rsma.2020.101467

Chapter 5

Stability over Time of the Quality of the Sulfurous Groundwater from the Deep Aquifer System That Supplies the Longroiva Medical Spa (Portugal)

Pedro Jorge Coelho Ferreira, Luis Manuel Ferreira-Gomes and Alcino Sousa Oliveira

Abstract

The sulfurous groundwater of Longroiva Medical Spa, having a use that comes from a long time ago, was the subject of several studies and research. In the last 50 years, its quality has been tested at the physical-chemical and microbiological levels in the different abstractions, namely, in the Classic Spring and the Well TD1, currently deactivated, and with greater detail in the current abstraction, Well AC1A. It was based on detailed studies, namely, systematic physical-chemical and microbiological analyses, that the sulfurous water from Well AC1A was legalized as natural mineral water; this situation made possible major investments in the place, such as the construction of the new Longroiva Medical Spa, among others. In this chapter, the stability of water quality over time is analyzed, based on studies resulting from the application of classical and multivariate statistical methods as well as the results of studies on the microbiome of these waters. The main conclusions are presented regarding the vulnerability of the aquifer and the exploration plan.

Keywords: sulfurous groundwater, natural mineral water, stability in quality, fissured aquifer, vulnerability, medical spa

1. Introduction

The sulfurous waters of Longroiva, associated with very extensive fractures of the Vilariça Fault family, have a very old tradition. Their first allusion appears in a document of the Order of Christ, dated 1570 [1]; later, they were referenced in Aquilégio Medicinal by Fonseca Henriques, the first inventory of medicinal waters in Portugal, published in 1726 [2] and later mentioned and studied by several authors, such as Almeida and Almeida [3], with the elaboration of the Hydrological Inventory of Portugal. The studies carried out by Ferreira-Gomes [4, 5] finally allowed their classification as a hydromineral and also a geothermal resource, with high potential for wealth creation, in the domains of classical thermalism, leisure, and well-being; in tourism; and with regard to geothermal use, as they are naturally hot water (47°C). The concession of that resource was attributed to the Municipality of Meda with the Exploration Contract signed on October 29, 2004, between the Municipality of Meda and the Portuguese State, having been published in the Portuguese Decree of January 26, 2005 [6].

Since 1970, these waters have been tested at physical-chemical and microbiological levels; however, it is since their legalization (2004) that they have come to comply with very demanding quality-control criteria. This control is carried out at several levels: (i) water is sampled on a weekly basis to carry out microbiological analyses (viable microorganisms at 37°C, viable microorganisms at 22°C, total coliforms, fecal coliforms, escherichia coli, fecal enterococci, pseudomonas aeruginosa, and spores of anaerobic sulfite-reducing bacteria); (ii) two to three annual harvests for physical-chemical analyses of the main chemical elements, namely, the global parameters, anions, and cations that represent the type of water in the aquifer being explored; (iii) complete physical-chemical analyses are carried out every 5 years for special control of trace chemical elements (metals, organic, and radiological), namely, those that may occur due to anthropogenic contamination.

Systematic analytical control of these waters makes it possible to assess their quality over time, detect possible contamination, and contribute to greater knowledge of the geohydraulic model of the aquifer system. In this sense, classical and multivariate statistical methods were applied to the results of the physical-chemical analyses collected during the licensing period for sulfurous water from the AC1A well (19 consecutive months) and to the samples obtained for quality control, over the last 20 years. Subsequently, a comparative analysis was executed with the results obtained in analyses carried out on water from other abstractions that served the Medical spa in Longroiva, prior to the AC1A well. At the microbiological level, the results of all the classic microbiology analyses carried out on the waters of Longroiva were analyzed, and the results of the study on the natural microbiology of these waters were analyzed.

2. Methodology

The methodology of this work is organized into 3 phases, namely:

1st phase - bibliographical research of published works with results of physical–chemical and microbiological analyses of the Longroiva waters, sampled in various abstractions that served the medical spa of Longroiva. We also used data on the microbiome of the waters of Longroiva, obtained in the Hidrogenoma project carried out by the Directorate-General for Energy and Geology [7].

2nd phase – analysis of results of physical-chemical and microbiological tests on water from well AC1A between 2014 and 2020.

3rd phase - treatment of data from physical-chemical and microbiological analyses of the Longroiva waters, in order to demonstrate the stability of these waters.

The analytical methods used in carrying out the analyses are generally in line with those presented in Directive 80/778/EEC [8].

3. Geographical, geomorphological, and geological framework

In the extreme west of Europe, located in the northeastern part of mainland Portugal, is located Vila de Longroiva, where the sulfurous waters that are the subject of study in this work occur; it is part of the Municipality of Meda, in the District of Guarda (**Figure 1**).

With regard to the geomorphological framework, in terms of relief, the region fits into a transition zone between the Central Plateaus, where flattened surfaces dominate, associated with slopes with steep slopes, and the plateau area of the surface of the Iberian Meseta [11]. This separation is materialized by the tectonic accident of the Vilariça fault (Bragança – Manteigas) [11], which compartmentalizes this area into two large blocks; this fault is characterized by being a left strike-slip deformation structure, with about 5.5 km of horizontal displacement that triggered a parallel fracturing in a range of 0.5 to 1 km wide, which, with the unevenness of the extreme blocks and the subsiding of the central block, originated the graben of Longroiva [12]. Locally and with interest for the current work, it is worth mentioning that at the geomorphological level, a mountain range with an average altitude of 750 m (granitic plateau, about 35 km long and 4 km wide) develops from Trancoso, in the NNE-SSW direction along the major axis, throughout the study area (**Figure 2**). In the eastern part, there is a vast plain, which reaches an altitude of 350 m. The drainage network that starts along the central mountain range (ridge line) is of the dendritic type and develops toward the Northeast, feeding the Massueime and Centieira streams toward the Côa river, and toward the Northwest, feeding the Teja stream, both tributaries on the left bank of the Douro River.

The area under study is part of the Ancient Massif, which corresponds to the western sector of the morphostructural unit of the Iberian Peninsula, called Hercynian basement [15], more specifically in the Central Iberian Zone, dominated by

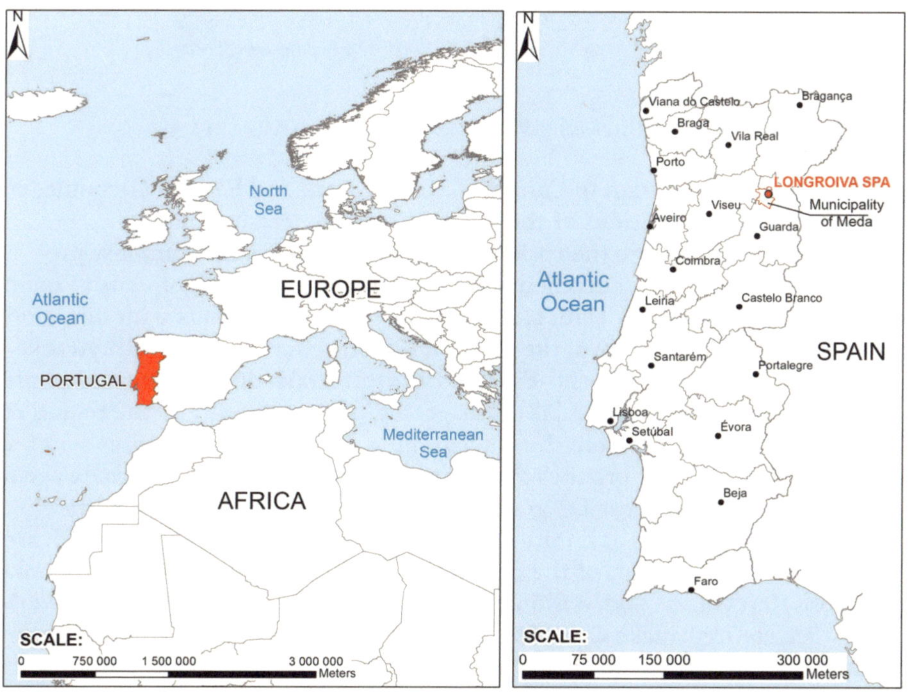

Figure 1.
Location of Longroiva Medical Spa [9, 10].

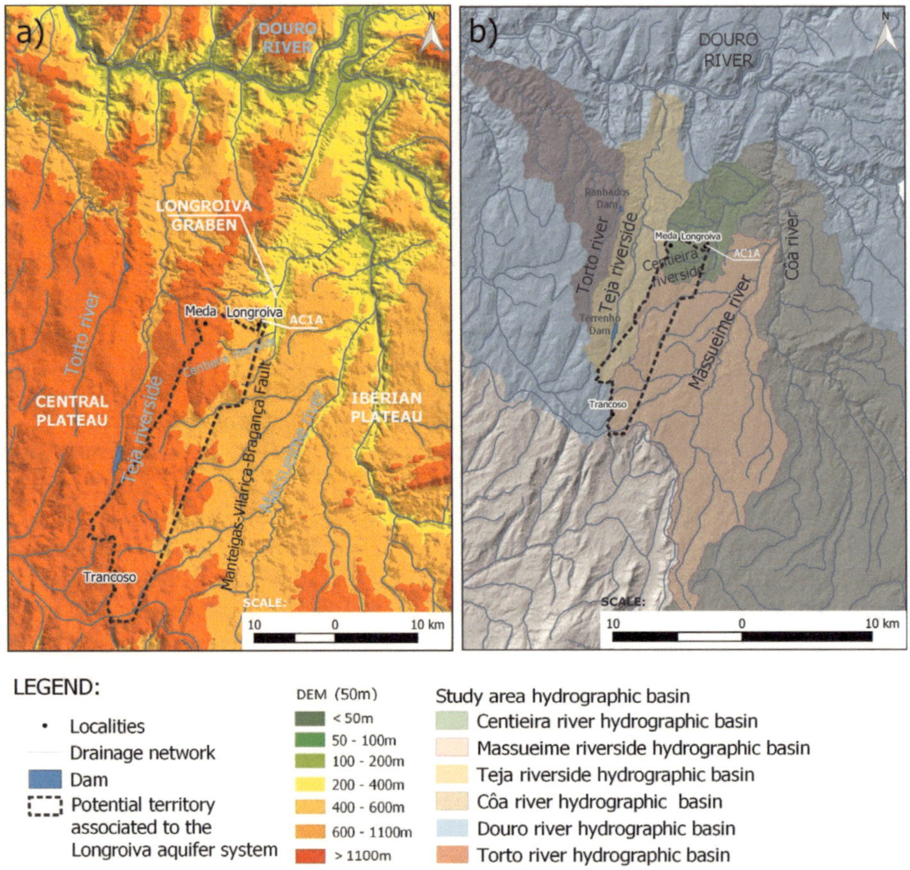

Figure 2.
Geomorphological framework: (a) Hypsometry of the region; (b) Hydrographic basin [13, 14].

formations from Precambrian to Cambrian, Ordovician, and Permo- Carbonic, covered by more recent formations of the Tertiary and Quaternary [12].

The oldest formations are the rocks of the Schist-Greywacke Complex, pre-Ordovician [16] (**Figure 3**); detached from those rocks are the alignments of quartzite crests from the Lower Ordovician, residual reliefs of syncline folds with direction E-W to ENE-WSW [19]; in turn, the granitoids instructed the Schist-Greywacke Complex in the Lamego-Penedono-Escalhão antiform axis, during the third phase of the Hercynian deformation (D3) [12]. The geological cartography of the region represents the sin-D3 granites, from the leucogranites and two-mica granites series as well as the series of biotitic granites and granodiorites, with an approximate age of 320–310 Ma, and the late-post-D3 granites, from series of biotitic-moscovitic granites as well as the series of biotitic granites and associated basic rocks, with around 310–290 Ma, the main groups of the Batholith Granitic of Beiras [20]. Throughout the late and post-Hercynic period, a filonian procession was installed that intruded that entire set (Schist-Greywacke Complex and granitoids) through distensive fractures, consisting of granite/riolitic porphyry dykes, microgabbros, alkaline basalts, masses pegmatitic or aplite-pegmatitic, and quartz veins [12]. Covering the oldest formations, there are the Cenozoic cover formations, which in the study area have greater importance, the Vilariça arkoses.

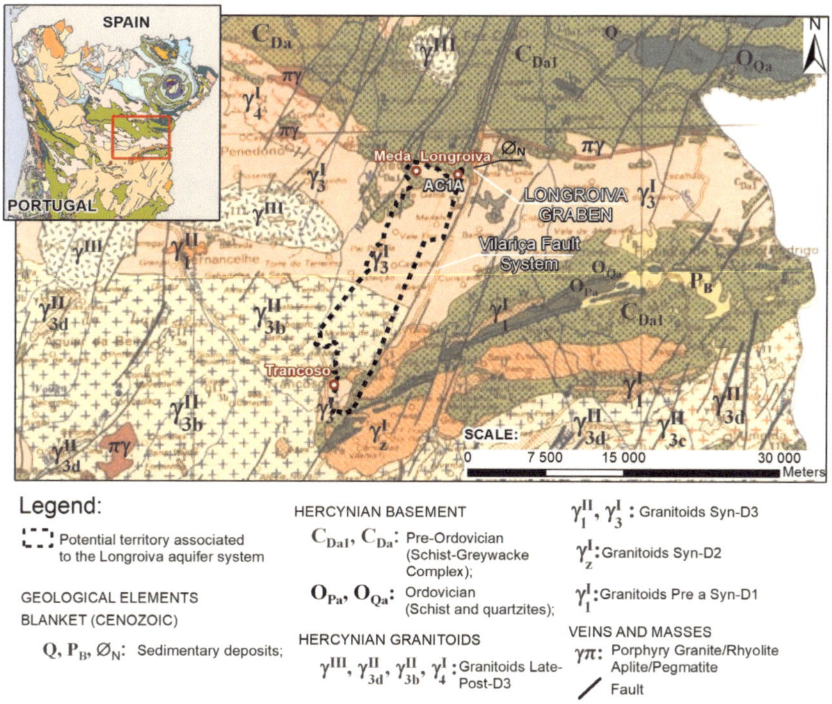

Figure 3.
Geological framework of the study area, adapted from [17, 18].

4. Hydrogeological aspects

The geohydraulic model of the Longroiva water aquifer system was initially advanced by Ferreira Gomes [4], during the works for the licensing of the water under study as a resource to be used in the Medical Spa of Longroiva, and later by Coelho Ferreira [21], the result of several studies developed within the scope of the PhD thesis, as shown in **Figure 4**.

In general, water of meteoric origin infiltrates the granite massif, since it is very altered and fractured, at altitudes between 770 and 809 m, along the granite plateau that extends from Trancoso to Meda. In the first phase, in the less extensive and more superficial circuits, the water will percolate in depth toward the thalwegs, emerging in some springs resulting from geological traps and forming small punctual aquifers dependent on the alteration and fracturing of the rock mass, of the free or phreatic type, which generally do not exceed 100 meters in depth. During these paths, water in a global way, due to the topography of the land, circulates inside the granite massif from west to east, until it reaches the faults in the NNE-SSW direction, the first barrier that conditions the circulation of water due to the difficulty that it has in crossing it. Thus, as a circulation corridor, the waters percolate along this fault from south to north; however, along the way, the intersections with the ENE-WSW to E-W family of faults allow the water to pass to other fracture corridors with NNE-SSW direction, east of the previous one.

In its downward underground path, it reaches the reservoir at a depth of 2000 m, reaching temperatures of around 115.4°C. After going through an extensive and deep

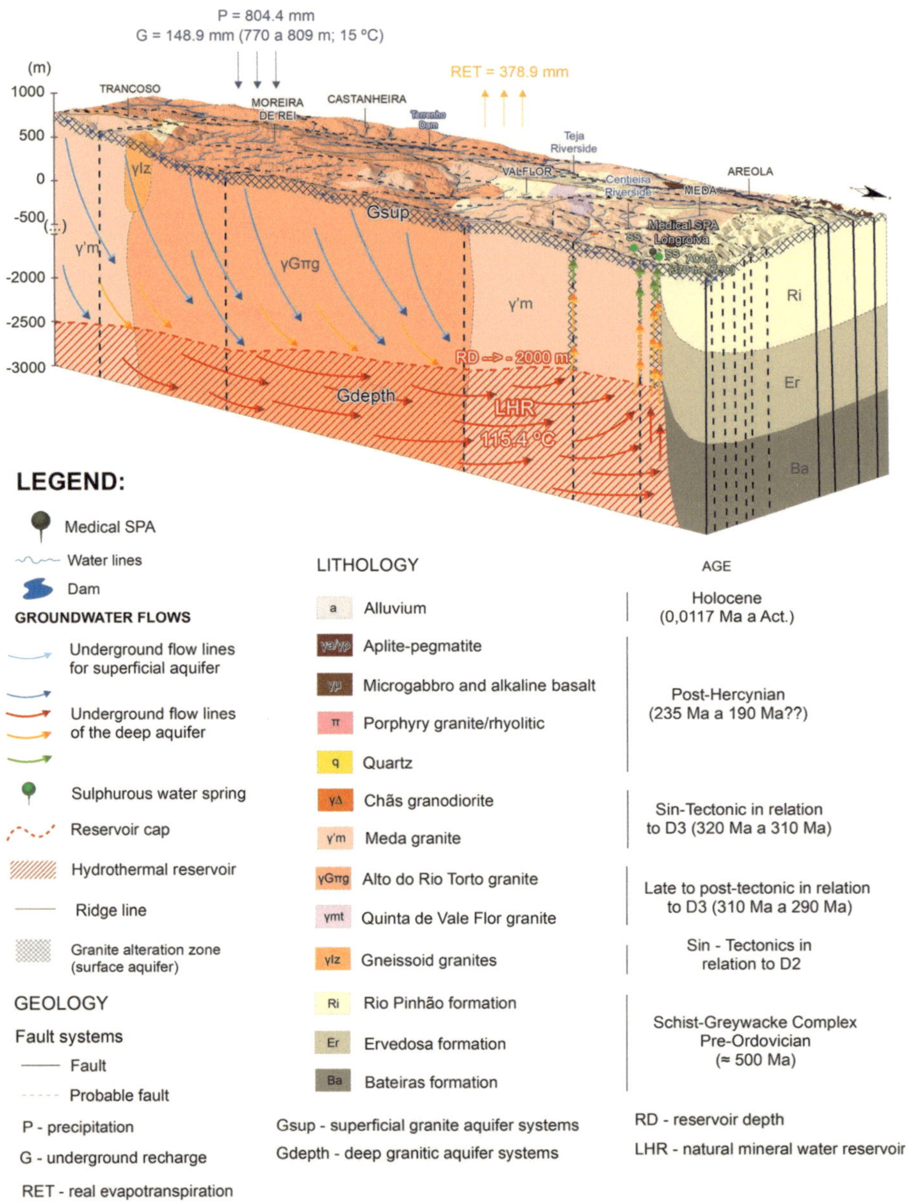

Figure 4.
Block diagram of the conceptual geohydraulic model of the sulfurous groundwater (natural mineral water) that supplies Longroiva Medical Spa [21].

underground circuit, where chemical alterations occur, resulting from water-rock-gas interaction processes and eventually even with a microbiological contribution, which give them unique and special characteristics, these waters emerge together with the contact of the granitoids with the metasedimentary formations (schists), which act as a barrier to their percolation. The emergence is further enhanced by hydraulic loads developed in the aquifer system and by the temperature reached in depth. Regarding discharges from the deep aquifer system, the schist/granite contact is clearly a barrier to the percolation of these waters in depth, and it is understood that, probably, the combination between the main families of faults, with global directions NNE-SSW and NW-SE, promote openings that allow their ascension.

The granitic massif next to the Longroiva Medical Spa constitutes two distinct aquifer systems: (i) the superficial granitic aquifer systems: of the free type, with common water, in the most superficial zone, around 100 m, characterized by not being very productive, and their recharging area in close proximity; (ii) deep granitic aquifer systems: of the confined type, with sulfurous water, and with a recharge area in a very distant zone. In previous studies [22], the hydraulic parameters for the sulfurous water-aquifer system were presented according to the following: hydraulic conductivity, $k = 0.4$ m/day; transmissivity, $T = 5.6\ m^2$/day; and storage coefficient, $S = 2.4 \times 10^{-3}$.

The abstraction is constituted by the AC1A well, which is a vertical well, 211.7 meters deep, and it should be noted that it is piped with stainless steel from the surface to 104.3 m deep and cement in its annular space. From that depth down, the well is not lined, and the water is captured between 104.3 and 211.7 m deep. The maximum flow rate for operation is 6.2 L/s, which is the value obtained in flowing artesian, with the resulting level at the mouth of the well, 0.6 m above the height of the natural terrain. The maximum piezometric level was recorded at 12 m above the natural ground level, when the hole is not delivering any flow, that is, with the faucet closed. It should be noted that as the piezometric operating level is lower (0.6 m) above the topographic surface, the potential for infiltration of contaminated fluids into the massif is consequently reduced.

5. Vulnerability and risk of aquifer system contamination

The vulnerability of groundwater to pollution can be defined, according to Van Duijvenbooden and Van Waegeningh [23], as “the sensitivity of groundwater quality to an imposed contaminant load, which is determined by the intrinsic characteristics of the aquifer”.

In the natural discharge zone, that is, in the area surrounding the Longroiva Medical Spa, Ferreira Gomes [24] established the vulnerability of the various geological units, based on the DRASTIC method [25] and some adaptations adjusted to the confined aquifer systems taking into account the abstraction site [24], resulting in the vulnerability map presented in **Figure 5**. From the same, it can be observed in particular that the lands in the AC1A well area present high to very high vulnerability, orienting that it will be necessary to be very careful with anthropic actions that may exist in that area as they could lead to contamination of the aquifer system.

Figure 5 also shows the Protection Perimeter for the waters of Longroiva, published in the Portuguese Decree [27], with the Extended Zone established to safeguard recharge areas [24]; the Intermediate Zone, generically corresponding to the areas close of the sulfurous water-aquifer system with considerable vulnerability; and the Immediate Zone, corresponding to the area with the greatest potential for natural discharge and where the AC1A well is located, being in particular an area to minimize the potential for water contamination in the abstraction itself.

In the area under study, two types of sources of pollution were considered [24]: risks associated with urban areas and risks associated with rural areas.

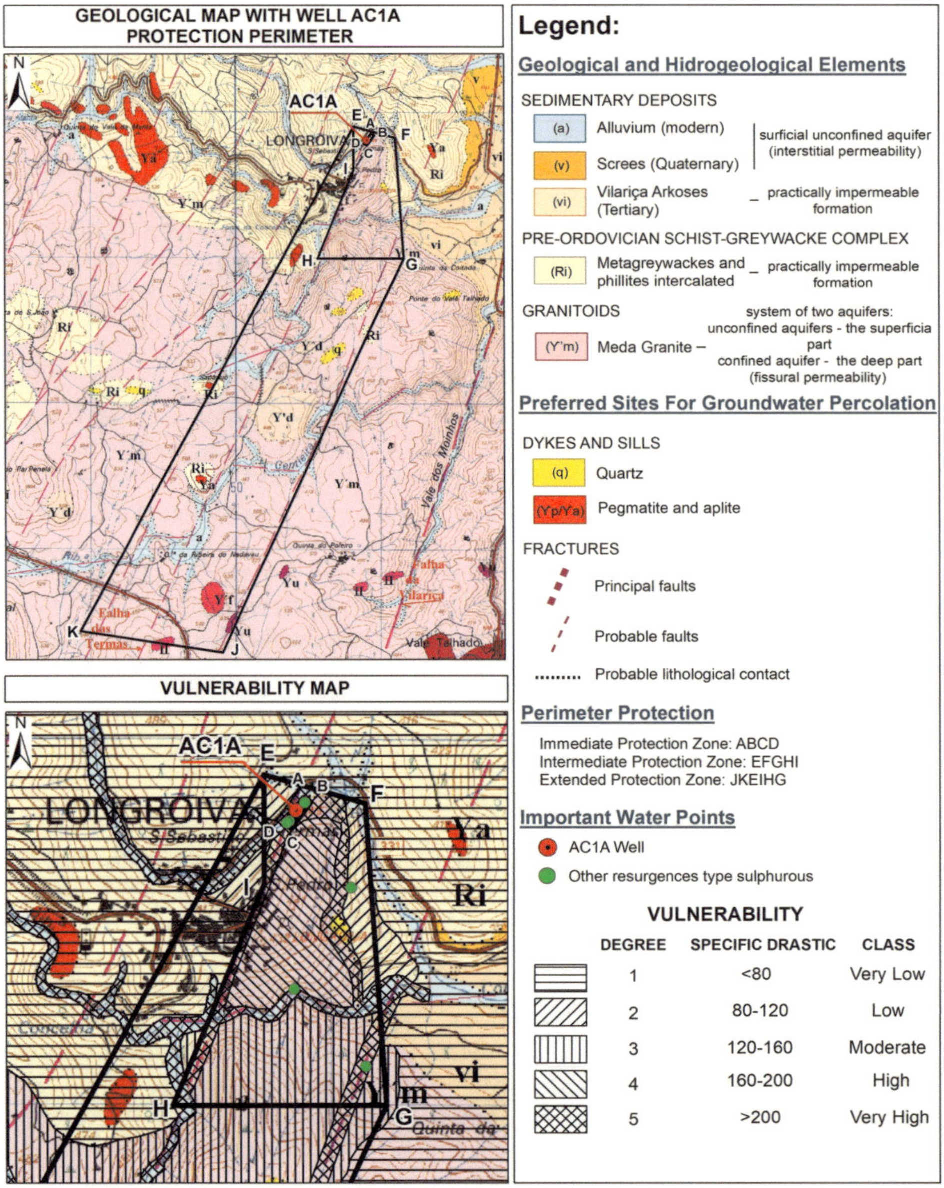

Figure 5.
Perimeter protection of groundwater extraction of Longroiva Medical Spa and vulnerability of the surrounding land (from [26]).

The risks associated with the urban area are linked to urban management, domestic activities, animal husbandry, and small industries, among others. Next to the Longroiva Medical Spa and in the surrounding areas, the following situations stand out: (i) punctual pits and in particular the general WWTP of the village, about 150 m downstream of Well AC1A; (ii) potential leaks from the sewer systems; (iii) family chicken coops that exist in the village; (iv) small industrial activity (mechanical workshop) that, despite being around 650 m from the Medical Spa, is located close to the Concelha riverside, with the land having high vulnerability; (v) the Longroiva cemetery, which should be treated with some care; (vi) the rural hotel of Longroiva that was built in the immediate area of the protection perimeter, taking advantage of the old thermal spa building; although with the works, there were positive actions of decontamination of the land next to the Classical Spring, and the anthropic pressure next to the Well AC1A increased, due to actions inherent to the hotel's activities.

The risks associated with rural areas are essentially due to intensive agricultural cultivation and use of fertilizers and pesticides. The use of these chemicals should be considered in places close to the Medical Spa and completely eliminated in the immediate and intermediate protection zones.

6. The water quality of the Longroiva Medical Spa aquifer system

6.1 Physical-chemical stability

According to several authors [4, 22], the water from the Longroiva Medical Spa, having a use that goes back a long time, has been the subject of several analyses and research. Of the studies or reports available at that time, the oldest analysis dates from 1970 [28], in water collected from the Classical Spring (NCL). The following analysis dates from 1988 [22]; in this case, the water sample had already been collected from Well TD1, which, however, had been built in order to improve the conditions for capturing and increasing the flow, with the intention of trying to legalize the Medical Spa. Similar results were obtained in the analyses carried out later over time (1988 to 1997), orienting for the water to always have the same origin and the same physical-chemical standard. In any case, the fact that the analysis in 1970 showed significant arsenic, as well as iron, copper and zinc, elements that are not part of the physical-chemical procession of the AC1A well water should be highlighted. Those elements with a high probability originated from the superficial aquifer system, which in the area was very contaminated. The contamination problem was overcome with the construction of the AC1A well, in which stainless steel tubing was carried out to a depth of 104.3 m and cementation in its annular space and also with the neutralization of all nearby springs and old wells, with cement grout injections [22]. **Figure 6** shows the location of the abstractions that served Longroiva Medical Spa and that were referred to earlier.

A study on the physical-chemical stability was carried out by analyzing the results obtained during the legalization phase of Well AC1A, in 1999/2000. The physical-chemical analyses were carried out in the laboratory of the former IGM (Geological and Mining Institute). The stability that the water showed then made it possible for it to be classified as natural mineral water, which could be used for medicinal purposes in the Longroiva Medical Spa, if the medico-hydrological study to be carried out was guided in this direction. It should be noted that 19 physical-chemical analyses were carried out, one per month, for 19 consecutive months. The basic statistical elements of the results of the analyzed parameters can be seen in **Table 1**, where it is verified that the relative standard deviation (RSD) of the total mineralization is only 1.92% and, in general, for the ionic parameters, it rarely exceeds 10%.

By projecting the average values relative to the analyses that were carried out in the water of the NCL, of Well TD1, as well as in the Well AC1A during the legalization phase in the Piper and Stiff diagram (**Figure 7**), it is verified that, without a doubt, we are facing the same water, although the analyses were carried out at different times and different laboratories.

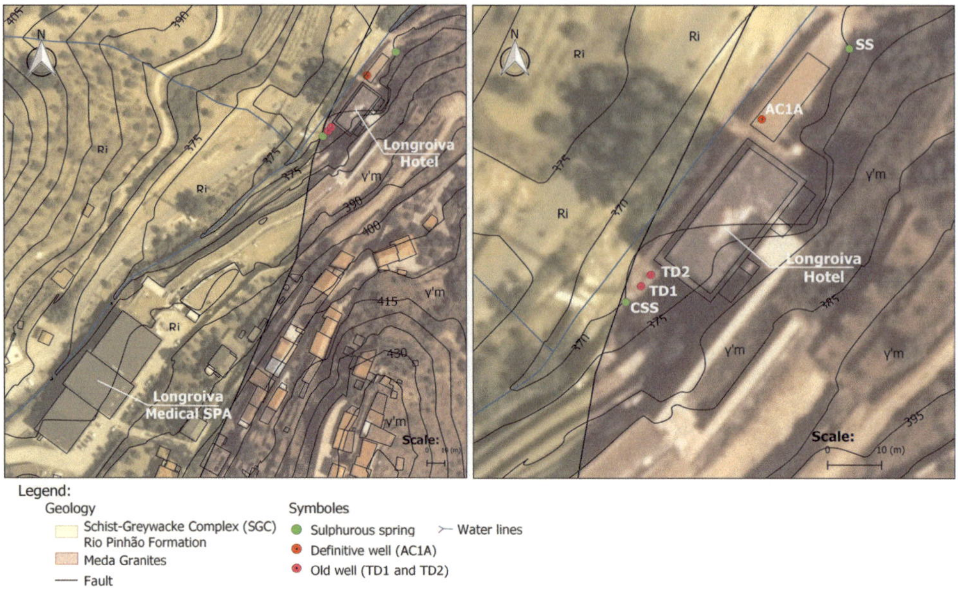

Figure 6.
Location plan of the abstractions that served the Longroiva Medical Spa [12, 29, 30].

Parameters		Units	Nr.	Min	Ave	Max	SD	RSD (%)
pH		—	19	8.71	8.86	8.93	0.05	0.55
Conductivity		$\mu S.cm^{-1}$	19	515.00	578.63	728.00	49.31	8.52
Total sulfuration		(em I_2 0.01 N) – mL/L	19	31.20	44.83	49.70	3.88	8.66
Alkalinity		$mg(CaCO_3)/L$	19	150.00	154.11	165.00	3.93	2.55
Total hardness		$mg(CaCO_3)/L$	19	6.50	7.05	8.00	0.36	5.10
Total CO_2		$mmolCO_2/L$	19	2.50	2.64	2.77	0.08	2.98
Silica - SiO_2		mg/L	19	58.30	64.50	71.10	2.92	4.52
Dry residue (at 180°C)		mg/L	19	375.00	395.26	407.00	8.27	2.09
Total mineralization		mg/L	19	441.00	462.16	474.00	8.85	1.92
Anions	HCO_3^-	mg/L	19	146.00	154.47	162.00	4.71	3.05
	Cl^-	mg/L	19	44.00	45.97	48.30	1.07	2.33
	SO_4^{2-}	mg/L	19	10.00	11.34	12.70	0.76	6.69
	F^-	mg/L	19	22.20	23.63	25.80	0.80	3.40
	CO_3^{2-}	mg/L	19	4.80	6.49	7.50	0.63	9.77
	NO_3^-	mg/L	19	<0.12	—	<0.38	—	—
	NO_2^-	mg/L	18	<0.01	—	<0.05	—	—
	HS^-	mg/L	19	5.10	7.38	8.20	0.65	8.77
	$H_3SiO_4^-$	mg/L	19	9.50	12.97	15.30	1.20	9.27
Cations	Na^+	mg/L	19	122.00	126.37	132.00	2.61	2.06
	Ca^{2+}	mg/L	19	2.60	2.73	3.00	0.13	4.73

Parameters	Units	Nr.	Min	Ave	Max	SD	RSD (%)
K^+	mg/L	19	4.10	4.57	5.30	0.25	5.40
Mg^{2+}	mg/L	19	<0.03	—	<0.10	—	—
Li^+	mg/L	19	0.59	0.77	0.88	0.08	9.90
NH_4^+	mg/L	19	0.42	0.53	0.65	0.06	10.75

Note: Nr.- number of samples, min- minimum, ave. - average, max - maximum, SD - standard deviation, RSD - SD relative.

Table 1.
Results in statistical terms of the main physical-chemical parameters of the water from AC1A well, during the licensing period 1999–2000 [22].

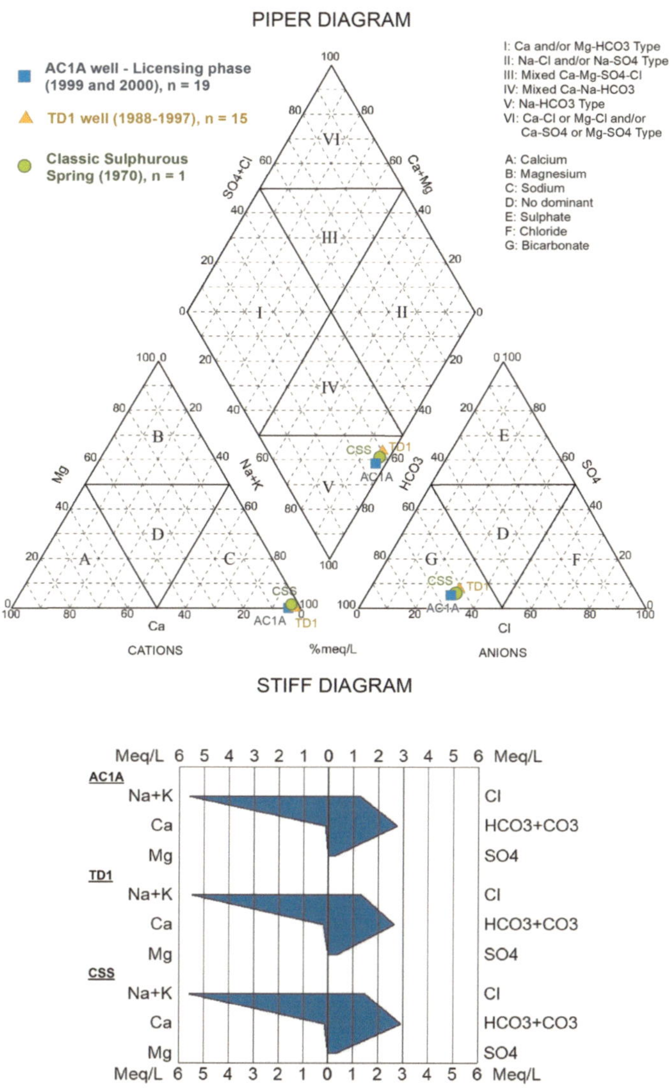

Figure 7.
Diagrams of Piper and Stiff, of the waters from the abstractions that served the Longroiva Medical Spa during the period 1970–2000.

After licensing the resource, the waters of Longroiva were subjected to systematic control, generally two to four analyses per year until today. Analyses during this period were always carried out by the Analysis Laboratory of the Instituto Superior Técnico in Lisbon. In order to be able to conclude the physical-chemical stability of these sulfurous waters, a detailed study was carried out by Coelho Ferreira *et al.* [26], which is now complemented. In **Table 2** and in **Figures 8–11**, the basic statistical results of the same parameters surveyed in that period are presented.

When analyzing the statistical results, it is verified that most of the parameters present excellent stability with RSD inferior to 10%, including the fact that it is a long period of research (20 years), and the main global parameters such as the total mineralization, the dry residue, the pH, the conductivity, and the silica in the non-ionized form present RSD values lower than 5%, a situation that leads to a great physical-chemical stability of the water; still, in the overall parameters, there is the

Parameters		Units	Nr.	Min	Ave	Max	SD	RSD (%)
pH		—	56	8.60	8.80	8.96	0.09	0.99
Conductivity		$\mu S.cm^{-1}$	56	527.00	541.42	688.00	20.97	3.87
Total sulfuration		(em I_2 0.01 N) - mL/L	56	34.00	43.73	53.00	4.46	10.20
Alkalinity		$mg(CaCO_3)/L$	55	138.00	149.93	155.00	3.79	2.53
Total hardness		$mg(CaCO_3)/L$	56	4.70	6.24	8.00	0.71	11.32
Silica - SiO_2		mg/L	56	59.00	65.41	73.00	3.24	4.95
Dry residue (at 180°C)		mg/L	56	367.00	383.04	398.00	6.79	1.77
Total mineralization		mg/L	56	437.50	456.08	475.00	7.06	1.55
Anions	HCO_3^-	mg/L	56	134.00	149.66	160.00	4.65	3.11
	Cl^-	mg/L	56	41.40	45.79	51.00	1.61	3.53
	SO_4^{2-}	mg/L	55	11.40	13.94	20.00	1.95	13.99
	F^-	mg/L	56	22.00	23.79	26.00	0.73	3.06
	CO_3^{2-}	mg/L	56	2.40	6.71	9.70	1.25	18.60
	NO_3^-	mg/L	56	<0.03	—	<0.03	—	—
	NO_2^-	mg/L	55	<0.005	—	<0.01	—	—
	HS^-	mg/L	54	5.70	7.21	9.00	0.72	10.06
	$H_3SiO_4^-$	mg/L	55	6.20	9.94	14.00	1.76	17.68
Cations	Na^+	mg/L	56	119.00	125.05	134.00	2.97	2.37
	Ca^{2+}	mg/L	56	1.90	2.49	3.10	0.25	10.21
	K^+	mg/L	56	3.70	4.84	8.70	0.94	19.37
	Mg^{2+}	mg/L	56	<0.1	—	<0.1	—	—
	Li^+	mg/L	56	0.44	0.73	0.80	0.05	6.66
	NH_4^+	mg/L	56	0.23	0.67	1.00	0.10	14.82

Note: Nr.- number of samples, min- minimum, ave. - average, max - maximum, SD - standard deviation, RSD - SD relative.

Table 2.
Results in statistical terms of the main physical-chemical parameters of the water from Well AC1A, during the control period (2001–2020).

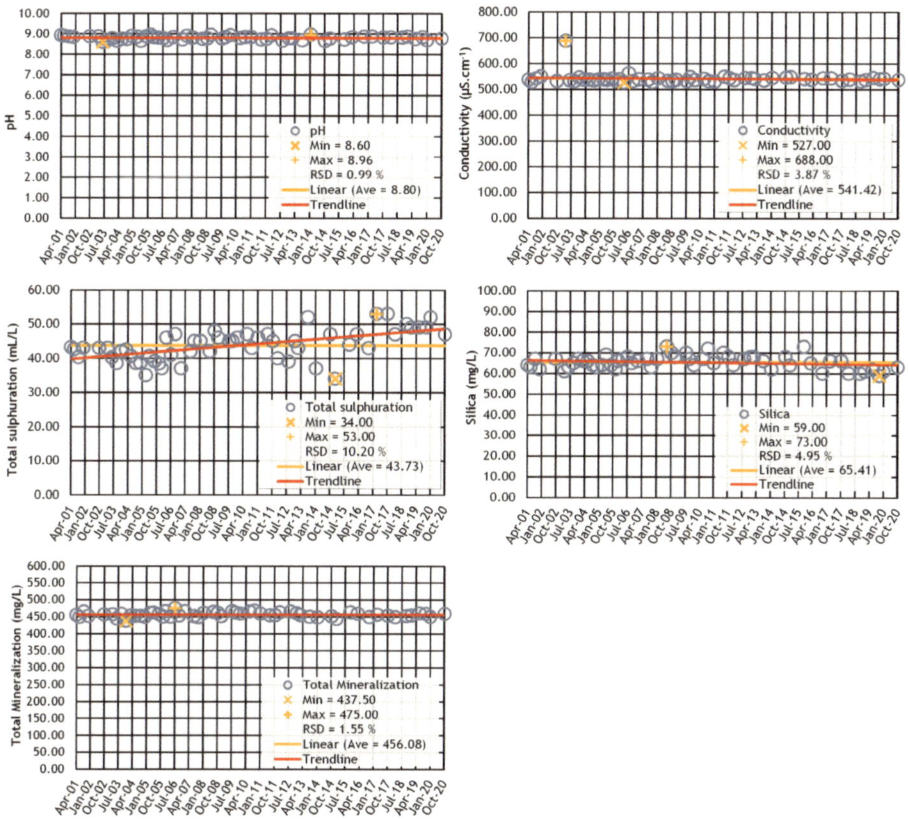

Figure 8.
Evolution over time of the physical-chemical parameters, in terms of the global parameters, of the water from Well AC1A, during the control period (2001–2020).

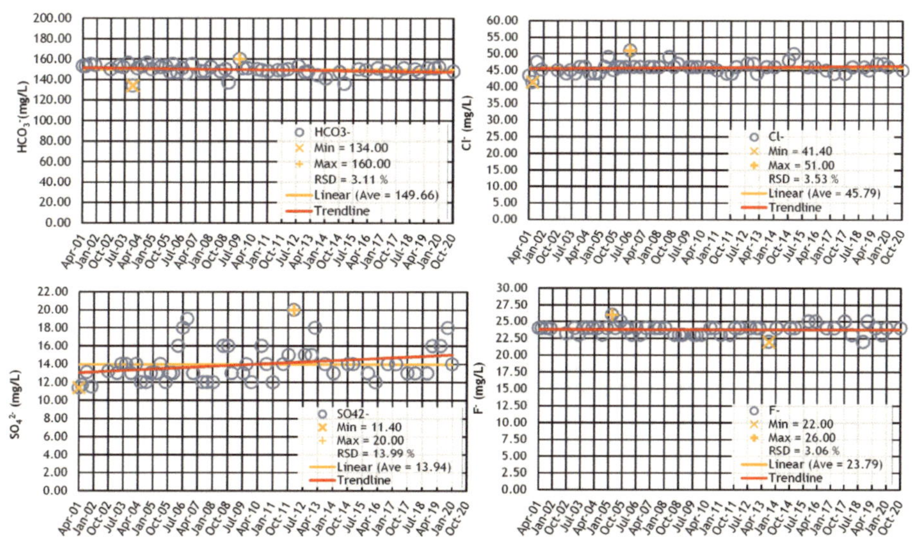

Figure 9.
Evolution over time of the physical-chemical parameters, in terms of the majority component - anions, of the water from Well AC1A, during the control period (2001–2020) – Sheet 1/2.

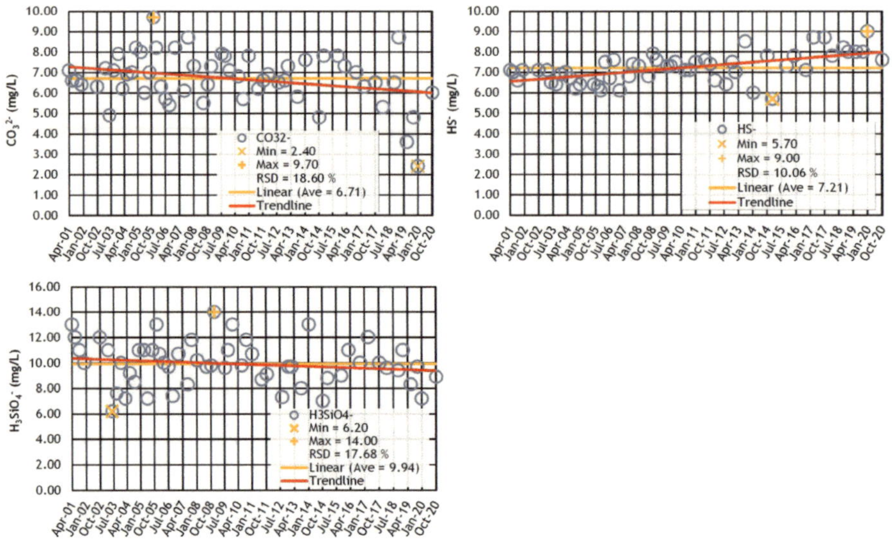

Figure 10.
Evolution over time of the physical-chemical parameters, in terms of the majority component - anions, of the water from Well AC1A, during the control period (2001–2020) – Sheet 2/2.

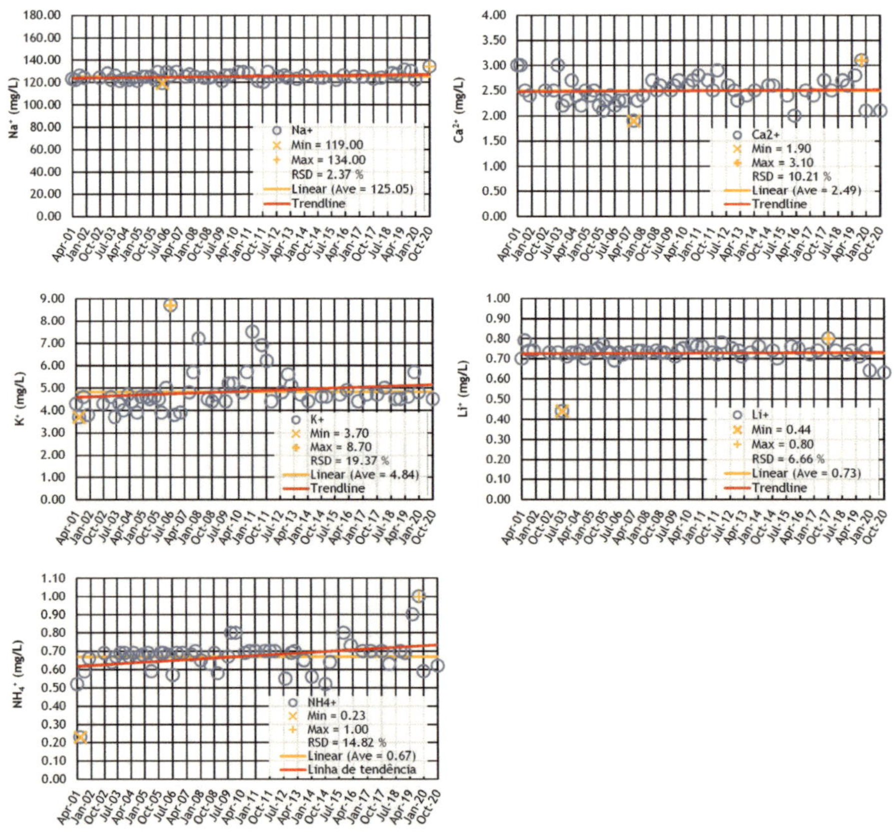

Figure 11.
Evolution over time of the physical-chemical parameters, in terms of the majority component - cations, of the water from Well AC1A, during the control period (2001–2020).

particularity of the total sulfuration showing a slight tendency to increase over time, despite the RSD still being in the order of magnitude of 10%. Regarding the ionic component, most of the parameters show statistically good stability; however, it is worth mentioning that some statistical trends increase or decrease over time, namely, in the cases of SO_4^{2-}, Hs^-, NH_4^+, CO_3^{2-}, $H_3SiO_4^-$, and K^+.

In **Figure 12**, the results obtained during the control phase are graphically presented and compared with the statistical values, minimum, average, and maximum, obtained in the legalization phase. It should be noted at this point that the fact that the results of the legalization phase were obtained in a laboratory different from the one used in the control phase may justify some small differences, resulting from various situations such as the handling of samples, as it should be noted that we are dealing with very small measurements, on the order of milligrams or parts per million.

Thus, when analyzing **Figure 12**, it is highlighted that for global parameters, most of the total values fall within the range between the minimum and maximum obtained in the legalization phase; exception is verified in the total hardness. With regard to ions, despite almost all of the parameters falling within the said fringe, there are some that are out of adjustment, namely: (i) in anions where SO_4^{2-} is essentially above, CO_3^{2-} oscillates above and below, and $H_3SiO_4^-$ is below; and (ii) in the cations Ca^{2+}, which is below, and K^+ and NH_4^+, which are essentially above.

A careful analysis of them leads to the mention that some of the deviations found may be essentially related to natural oscillations and even to the tuning of analytical techniques. Of note, the case of silicate ($H_3SiO_4^-$) as the example containing more

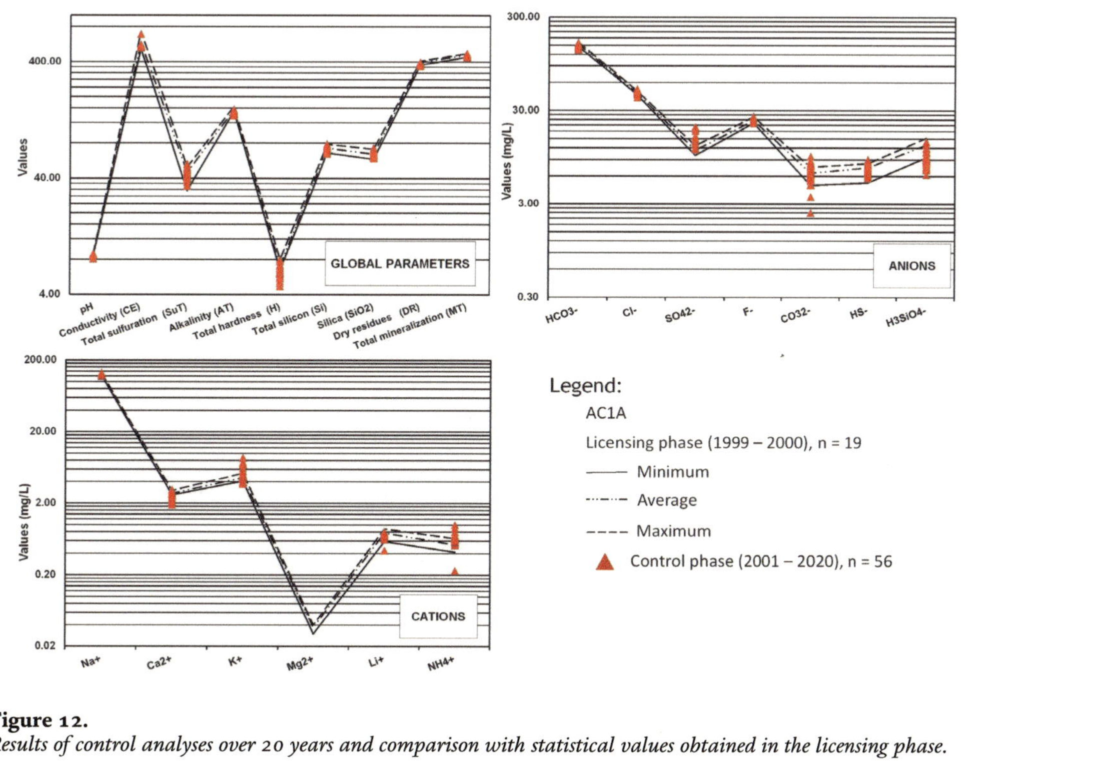

Figure 12.
Results of control analyses over 20 years and comparison with statistical values obtained in the licensing phase.

significant oscillations, it is also verified that the range of values in the legalization period (1999 and 2000) is above that in the control period (2001 to 2020), and consequently, the values obtained in the IGM laboratory are higher than those obtained in the IST laboratory.

In any case, this simple analysis leaves some doubts, and therefore, there is a need to continue the study with other more specific research techniques. Thus, mathematical techniques such as Principal Component Analysis (PCA) were used. To carry out the PCA on the results of the chemical analyses of water from the Well AC1A, two phases were considered: 1st phase corresponding to the legalization period (1999–2000); 2nd phase corresponding to the control period (2001–2020). For the 1st phase, the matrix of 19 samples was considered, corresponding to the 19 consecutive months during the years 1999 and 2000, with 17 variables (physical-chemical parameters). For the 2nd phase, the matrix of 56 samples was considered, corresponding to all existing analyses of Well AC1A, with 17 variables (physical-chemical parameters).

The multivariate statistical analysis was performed using the Andad software, version 7.12 [31]. The results obtained are presented in **Table 3**, where the eigenvalues (Vp), the explained variance for each axis (Vi), and the accumulated variance (Vc) are observed. The choice of the number of factor axes to retain was based, once again, on considering those with eigenvalues greater than 1.

In studies carried out by Coelho Ferreira *et al.* [26], the same exercise was carried out, with the control phase including analyses only up to 2013, and the PCA results of the two phases were similar. In the present work, the analyses up to 2020 were considered in the control phase, and the addition of this information allowed verifying that the results of the multivariate statistics (PCA) point to the existence of a trend of slight differences in the two phases. When analyzing the results of the PCA (**Table 3**),

Study phase	Axis	Variables explained by axis	V_p	V_i (%)	Vc (%)
1st phase: 1999–2000 (Licensing period)	1	+Si, $+SiO_2$, +RS, $+M_T$, $+Na^+$, $+HCO_3^-$	4.98	29.28	29.28
	2	$+NH_4^+$, $-Cl^-$, $+SO_4^{2-}$, $+F^-$, $-CO_3^{2-}$, -$H_3SiO_4^-$	2.76	16.25	45.53
	3	$+S_T$, $+HS^-$	2.06	12.11	57.64
	4	$-Li^+$, $+F^-$	1.74	10.25	67.89
	5	$-K^+$, $+SO_4^{2-}$	1.54	9.05	76.94
	6	$+Cl^-$	1.28	7.55	84.49
2nd phase: 2001–2020 (Control period)	1	-Si, $-SiO_2$, -RS, $-M_T$	2.99	17.60	17.60
	2	$+S_T$, $+HS^-$, $-F^-$, $-CO_3^-$	2.60	15.31	32.91
	3	$^-$ + Si, $+SiO_2$, $-Na^+$, $-HCO_3^-$	2.01	11.83	44.74
	4	$+Li^+$, $+H_3SiO_4^-$	1.66	9.78	54.52
	5	$+HCO_3^-$, $-Cl^-$	1.44	8.45	62.98
	6	$-Ca^{2+}$, $+SO_4^{2-}$	1.24	7.27	70.25
	7	—	1.08	6.37	76.62

Table 3.
Result of the PCA applied to the results of the physical-chemical analysis from Well AC1A waters.

namely, the values of the variables of the factorial axes as well as their projection in the first and second factorial plan, it is worth mentioning the following for the licensing phase:

- The first factor axis explains about 29.28% of the total variance; this factor is determined by the lithological/environmental and long temporal context where the water/rock interaction will determine the relatively high and complex mineralization of this water. Another aspect explained in this factorial axis is the hydrochemical facies of this water (sodium bicarbonate).

- The second factorial axis explains about 16.25% of the total variance; this factor can be explained by the presence of sulfur in the hydromineral system, which determines the presence of N in the form of NH_4^+, thus not relating this chemical species to anthropogenic phenomena. The NH_4^+ species is related to F^- and SO_4^{2-} and in opposition to the CO_3^{2-}, Cl^-, and $H_3SiO_4^-$ species.

- The third factorial axis explains about 12.11% of the total variance; the $S_T - HS^-$ association is strongly correlated with this factorial axis; this association shows the presence of sulfur in these waters, in its reduced forms.

In the case of the control phase, which includes physical-chemical analyses of these waters for 20 years, the results obtained and in comparison with the licensing phase were considered as follows:

- The first factor axis explains about 17.60% of the total variance; this factor remains, in relation to the licensing phase, as determined by the lithological/environmental and long temporal context where the water/rock interaction will determine the relatively high and complex mineralization of this water.

- The second factor axis explains about 15.31% of the total variance; in this factor axis, contrary to the licensing phase, the $S_T - HS^-$ association gains greater preponderance; this association, which shows the presence of sulfur in these waters in its reduced forms, has had an increasing tendency over the years, most likely due to the evolution of the entire system toward stability. In opposition to the previously mentioned association, CO_3^{2-} and F^- are found, with CO_3^{2-} showing a decreasing trend over the years.

- The third factorial axis explains about 11.83% of the total variance; this factorial axis reflects the hydrochemical facies of this water (sodium bicarbonate). The association between Na^+ and HCO_3^- occurs in opposition to the association between SiO_2 and Si.

To obtain a better understanding of these aspects, the graphs in **Figure 13** are presented, with the results available so far from all abstractions with sulfurous water that have already served the Longroiva Medical Spa. The referred graphs show the analytical results obtained in the Classical Spring in 1970 (n = 1) and in the Well TD1 during the period from 1988 to 1997 (n = 15), and these are compared with the statistical values, minimum, average, and maximum, obtained in the phase of legalization and control of Well AC1A, in the period between 1999 and 2020 (n = 75), which previously proved its physicochemical stability. When analyzing the results, it is

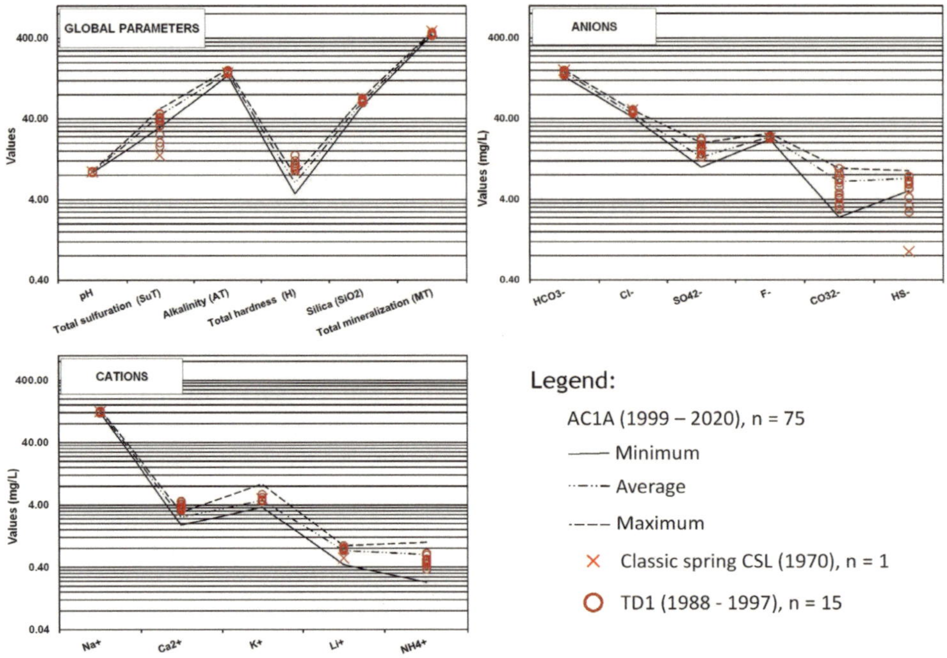

Figure 13.
Comparison of the results over time of the physical-chemical parameters, in terms of the global parameters and major elements (anions and cations), of the sulfurous water from the various abstractions of the Longroiva Medical Spa.

verified that most of the global parameters are within the range between the minimum and the maximum mentioned above, with the exception of total sulfuration (S_T) and hardness (H). In the case of ions, most of them are also within this range, with the exception of anions being SO_4^{2-} and HS^- and Ca^{2+} in cations. Once again, the physical-chemical stability of this water over the 50 years during which analyses were carried out is demonstrated. With regard to existing variations in some elements, they may probably be related, as previously mentioned, to natural oscillations and even to the tuning of analytical techniques.

Special control analyses of trace chemical elements (metals, organic, and radiological), namely, those that may occur due to anthropogenic contamination, are carried out less frequently. During the licensing and control phases, only three analyses of this type were carried out, in 2000, 2016, and 2021. The results obtained are shown in **Table 4**, and from the analysis carried out on them, the following should be mentioned:

- the trace elements that form part of the typical procession of these waters, B, Cs, Rb, Sr., and W show an increasing trend over time, with W standing out as the element that most increased its concentration.

	Parameteres	2000	2016	2021
Trace Elements (μg/L)	Ag	<0.5	<0.1	<0.1
	Al	<12.0	8.0	3.0
	As	<3.0	3.5	4.8
	B	288.0	300.0	304.0

DOI: http://dx.doi.org/10.5772/intechopen.109762

	Parameteres	2000	2016	2021
	Ba	107.0	30.0	<30.0
	Be	<1.0	0.6	0.5
	Bi	—	<10	<0.8
	Cd	<1.0	<1.0	<0.40
	Co	<6.0	<2.0	<2.0
	Cr	<6.0	<1.0	<1.0
	Cs	—	136.0	153.0
	Cu	<2.0	<2.0	<2.0
	Hg	—	<0.05	<0.05
	Fe	<3.0	<0.01	<0.010
	Mn	1.1	<5.0	<5.0
	Mo	<4.0	<5.0	<5.0
	Nb	<4.0	<1.0	<1.0
	Ni	—	<5.0	<5.0
	Pb	<6.0	<3.0	<3.0
	Rb	—	87.0	93.0
	Sb	<3.0	<1.0	<1.0
	Se	<3.0	<0.4	<0.4
	Sn	—	<5.0	<5.0
	Sr	88.0	90.0	96.0
	U	—	<0.1	<0.1
	V	<2.0	<10.0	<10
	W	82.0	153.0	316.0
	Y	<1.0	<0.5	<0.50
	Zn	<2.0	<50.0	<50.0
	Zr	—	<1.0	<1.0
Organic elements	Hydrocarbons (μg/L)	—	*	*
	Trihalomethanes (μg/L)	—	*	*
	Volatile Organic Compounds (μg/L)	—	*	*
	Polycyclic Aromatic Hydrocarbons (ng/L)	—	*	*
	Pesticides (μg/L)	—	*	*
Radiological elements (Bq/L)	Alpha total	—	<0.04	<0.04
	Beta total	0.329	0.2	0.14
	Radon	—	50.0	60.8
	^{226}Ra	0.029	—	—
	Tritium	—	—	<10.0

**value lower than detection limit.*

Table 4.
Result of trace elements obtained through the complete physical-chemical analysis of the Well AC1A water.

- In the case of elements that may reveal possible anthropogenic contamination, namely, metals and organic elements, there is no significant increase or trend that points to a situation of potential contamination of the hydromineral resource.

Although it was possible to highlight some aspects of the available analytical results, the incipient knowledge about these waters, as well as the reduced number of analyses over such a long time, does not allow reaching consistent conclusions, much less triggering preventive actions that mitigate the possible contamination of the mineral aquifer.

6.2 Microbiological elements

Mineral water, such as the water from Longroiva Medical Spa, in addition to its chemistry as an intrinsic and consequent characteristic of the hydrogeological circuit traversed, may also have a natural microbiome inherent in each water, in addition to having microorganisms that are not common to their normal characteristics and are even pathogenic; the latter are normally associated with anthropic actions and are considered here as research within the classical microbiology of groundwater.

6.2.1 Classical microbiology

Control in terms of classical microbiology in groundwater of the abstractions is directed toward research and quantification of: viable microorganisms – number of colonies at 36 ± 2°C, viable microorganisms - number of colonies at 22± 2°C, total coliforms, fecal coliforms, escherichia coli, fecal enterococci, pseudomonas aeruginosa, and spores of anaerobic sulfite-reducing bacteria.

These microorganisms, researched for either the licensing of new mineral waters or the use of mineral waters, in order to verify if they are in conditions of suitability in terms of public health, are defined in accordance with Law n.° 1220/2000.

Thus, research on classical microorganisms provides guidance on the quality of the terminal phase of the geohydraulic circuit and natural mineral water abstraction, being important for each water point (springs, wells, holes), as far as possible, to have a history of these aspects.

The Longroiva Medical Spa, as already mentioned, goes a long way back in time; however, in the past, there was no microbiological control. Meantime, the rules were changing, and for Longroiva Medical Spa to be licensed, it had to have its resource, which was initially obtained from the Classical Spring, with adequate microbiological results. This did not happen for many years, as the discharge zone of the geohydraulic circuit for the sulfurous water of Longroiva Medical Spa was contaminated; then, it evolved into the Well TD1 with 40 m of depth, and the situation ended up not changing. The results that it was possible to compile by Ferreira Gomes [32] attesting to such situations are presented in **Table 5**, showing that the water was systematically unsuitable, namely, total coliforms, sometimes fecal coliforms, fecal streptococci, and sulfite-reducing spore-forming anaerobes, plus excess total germs.

However, relatively recent works have evolved toward Well AC1A, capturing sulfurous water below a depth of 104 m, neutralizing some points of sulfurous water in the vicinity, namely, Well TD1, and the classic microbiological results were adequate in a distinguished way; **Table 6** shows the results obtained, from May 1999 to November 2000, during the legalization phase of Longroiva Medical Spa [22].

Parameters	MRV	MAV	Date (year/month)										
			69/07 (0)	90/07 (0)	97/07 (1)	97/08 (1)	97/09 (1)	97/10 (1)	97/11 (1)	97/12 (1)	98/01 (1)	98/02 (1)	98/03 (1)
1	5	20	76	10	70	>300	140	42	6	18	8	>300	40
2	20	100	—	90	50	>300	>300	120	30	22	3	>300	20
3	0	0	—	14	>80	>80	>80	>80	3	0	12	0	>80
4	0	0	—	0	0	0	0	0	0	0	12	0	0
5	0	0	—	0	13	50	15	1	0	0	0	0	0
6	0	0	—	0	0	0	0	0	0	0	0	0	0
7	0	0	—	0	>1	>1	>1	1	0	0	0	0	0

Research: 1 – Total germs (37°C, 24 h), 2 – Total germs (22°C, 72 h), 3 – Total coliforms, 4- Fecal coliforms, 5 – Fecal streptococci, 6 - Pseudomonas aeruginosa, 7 - Sulfite-reducing clostridia.
MRV - Maximum Recommended Value, MAV - Maximum Admissible Value.
(0)(1) Results of the water collected, respectively, in the Classic Spring and in the Well TD1.

Table 5.
Results of bacteriological analyses carried out on Classic Spring and TD1 Well water over the years [32].

Date	Research						
	1	2	3	4	5	6	7
04/05/99	1	1	0	0	0	0	0
09/06/99	2	4	0	0	0	0	0
20/07/99	1	2	0	0	0	0	0
24/08/99	1	1	0	0	0	0	0
07/09/99	2	3	0	0	0	0	0
19/10/99	1	1	0	0	0	0	0
10/11/99	2	1	0	0	0	0	0
07/12/99	1	4	0	0	0	0	0
24/01/00	1	0	0	0	0	0	0
08/02/00	2	1	0	0	0	0	0
14/03/00	1	2	0	0	0	0	0
04/04/00	1	3	0	0	0	0	0
02/05/00	1	1	0	0	0	0	0
19/06/00	1	2	0	0	0	0	0
06/09/00	0	0	0	0	0	0	0
26/09/00	0	0	0	0	0	0	0
10/10/00	0	0	0	0	0	0	0
21/11/00	1	1	0	0	0	0	0

Research: 1 – N.° of colonies per mL, at 37°C, 24 hours; 2 – N.° of colonies per mL, at 22°C, 72 hours; 3 – N.° of total coliforms per 250 mL; 4 – N.° of fecal coliforms per 250 mL; 5 - Count of fecal streptococci per 250 mL; 6 - Quantification of spores of anaerobic sulfite-reducing bacteria per 50 mL; 7 - Quantification of pseudomonas aeruginosa per 250 mL.

Table 6.
Results of bacteriological analyses carried out on the water from the Well AC1A for the legalization of Longroiva Medical Spa [22].

Year	N.° of Analysis	Research							
		1	2	3	4	5	6	7	8
2001	17	17 < RV	17 < RV	17 < LV	—	17 < LV	17 < LV	17 < LV	17 < LV
2002	31	29 < RV – 2 > RV	30 < RV – 1 > RV	31 < LV	—	31 < LV	31 < LV	31 < LV	31 < LV
2003	29	28 < RV – 1 > RV	28 < RV – 1 > RV	29 < LV	—	29 < LV	29 < LV	29 < LV	29 < LV
2004	22	20 < RV – 2 > RV	22< RV	22 < LV	—	22 < LV	22 < LV	22 < LV	22 < LV
2005	37	37< RV	37< RV	37 < LV	—	37 < LV	37 < LV	37 < LV	37 < LV
2006	34	34 < RV	34 < RV	34 < LV	—	34 < LV	34 < LV	34 < LV	34 < LV
2007	35	35 < RV	35 < RV	35 < LV	—	35 < LV	35 < LV	35 < LV	35 < LV
2008	36	35 < RV – 1 > RV	36 < RV	36 < LV	—	36 < LV	36 < LV	36 < LV	36 < LV
2009	35	34 < RV – 1 > RV	35 < RV	35 < LV	—	35 < LV	35 < LV	35 < LV	35 < LV
2010	34	34 < RV	34 < RV	34 < LV	—	34 < LV	34 < LV	34 < LV	34 < LV
2011	38	38 < RV	38 < RV	38 < LV	38 < LV	38 < LV	38 < LV	38 < LV	38 < LV
2012	29	29 < RV	29 < RV	29 < LV	29 < LV	29 < LV	29 < LV	29 < LV	29 < LV
2013	34	34 < RV	34 < RV	34 < LV	34 < LV	34 < LV	34 < LV	34 < LV	34 < LV
2014	40	35 < RV – 5 > RV	40 < RV	40 < LV	40 < LV	40 < LV	40 < LV	40 < LV	40 < LV
2015	46	45 < RV – 1 > RV	46 < RV	45 < LV – 1 > LV	46 < LV	46 < LV	46 < LV	46 < LV	46 < LV
2016	52	52 < RV	51 < RV – 1 > RV	52 < LV	52 < LV	52 < LV	52 < LV	52 < LV	52 < LV
2017	52	52 < RV	52 < RV	52 < LV	52 < LV	52 < LV	52 < LV	52 < LV	52 < LV
2018	52	52 < RV	52 < RV	52 < LV	52 < LV	52 < LV	52 < LV	52 < LV	52 < LV
2019	52	51 < RV – 1 > RV	51 < RV – 1 > RV	52 < LV	52 < LV	52 < LV	52 < LV	52 < LV	52 < LV
2020*	13	13 < RV	13 < RV	13 < LV	13 < LV	13 < LV	13 < LV	13 < LV	13 < LV

Research: 1 – Quantification of cultivable microorganisms at 37 °C, 24 hours (CFU/1 mL); 2 – Quantification of cultivable microorganisms at 22 °C, 72 hours (CFU/1 mL); 3 – Research and quantification of total coliforms (CFU/250 mL); 4 – Research and quantification of fecal coliforms (CFU/250 mL); 5 – Research and quantification of Escherichia Coli (CFU/250 mL); 6 – Research and quantification of fecal enterococci (CFU/250 mL); 7 – Research and quantification of Pseudomonas aeruginosa (CFU/250 mL); 8 – Research and quantification of spores of anaerobic sulfite-reducing bacteria (CFU/50 mL).
Law 1220/2000 - Recommended Value (RV); RV = 5 CFU/1 mL for research 1; RV = 20 CFU/1 mL for research 2. Limit Value - maximum admissible value (LV); LV = 0 CFU/1 mL for research 3, 4, 5, 6, 7, and 8.
**In 2020, due to the COVID-19 pandemic, the Hydromineral Resource Exploitation Plan was suspended as of March 17, in the thermalism aspect, resulting in only 13 microbiological analyses throughout the year.*

Table 7.
Results of the bacteriological analyses carried out on the water of the Well AC1A throughout the control phase, in a total of 718 analyses (compiled from records in technical reports of the spa's activities over time, held by the DGEG and the concessionaire (CMM)).

Following the legalization of medical spas and since they are in operation, it is usual to comply with an annual analytical control plan, determined by the licensing entity, in which the number of microbiological analyses to be carried out monthly is defined. In the case of Longroiva Medical Spa, over about 20 years, it fluctuated between 1 and 5 monthly analyses, a total of 718 (**Table 7**).

From the results obtained, it appears that most analyses of the Well AC1A waters are below the MAV in microorganisms cultivable at 37°C and 22°C, and for the remaining microorganisms analyzed, in 20 years, the detection of a CFU of total coliforms occurred only once, in 2015, during the works of the hotel. In this way, the quality of construction of the Well AC1A becomes evident, with the capture of the resource at a depth greater than 104 m and also the relationship between anthropic actions in the surroundings of the well.

6.2.2 Aspects about microbiome

Groundwater has a natural microbial community in its composition, which has been studied over the years; however, knowledge about these aspects is still considered incipient.

The Directorate General for Energy and Geology (DGEG) led the HIDROGENOMA project, during the period 2016 to 2019, which aimed to investigate the natural microbism of 80 Portuguese natural mineral waters, which included the sulfurous water from Longroiva Medical SPA [7]. According to the same authors, this project aimed to correlate scientific knowledge in the areas of geology, hydrogeochemistry, and microbiology and contribute to better management, exploitation, and enhancement of hydromineral resources. The results, among others, will be able to infer about the microbiological ecosystems of the aquifers crossed, giving guidance on the conditions of pressure, temperature, and degree of oxidation-reduction of the environment that will occur in them [7]; thus, it is understood that those elements will contribute to the knowledge of the hydrogeological models of each aquifer system.

The project methodology consisted of collecting four samples at the head of the abstractions, during the years 2017 and 2018, in spring and autumn, in the 80 natural mineral waters that were contemplated in this study [7]. Also, according to Lourenço and Pascoal [7], after each sampling phase, a genomic study of the waters was carried out in the laboratory.

Table 8 presents the results obtained for the various samples, in terms of the taxonomic composition of bacterial communities, based on taxonomic affiliation by class. Using the mean values per class, as shown by DGEG [34], **Figure 14** shows the distribution of the various classes identified; there are 7 most-frequent bacterial classes, with two dominant: Gammaproteobacteria (20.09%) and Betaproteobacteria

Class	%_Reads F1	%_Reads F3	%_Reads F5	%_Reads F7	Average	Standard Deviation
Gammaproteobacteria	56.07	8.46	7.66	8.17	20.09	23.989
Betaproteobacteria	34.99	2.80	31.00	10.41	19.80	15.635
Unclassified	1.90	29.76	8.27	29.00	17.23	14.269
Deltaproteobacteria	0.40	29.31	10.62	5.26	11.40	12.650
Nitrospira	—	7.51	21.33	4.49	11.11	8.979

Class	%_Reads F1	%_Reads F3	%_Reads F5	%_Reads F7	Average	Standard Deviation
Alphaproteobacteria	5.21	2.84	—	23.86	10.64	11.513
Clostridia	0.20	9.41	11.22	4.93	6.44	4.929
Deinococci	0.73	2.29	—	3.27	2.10	1.281
Actinobacteria	—	—	3.44	—	—	—
Flavobacteriia	0.20	—	—	—	—	—
Nostocophycideae	—	—	3.59	—	—	—

Table 8.
Comparative summary of the taxonomic composition of bacterial communities, based on taxonomic affiliation by class, for samples of sulfurous water from Well AC1A at Medical Spa Longroiva [33].

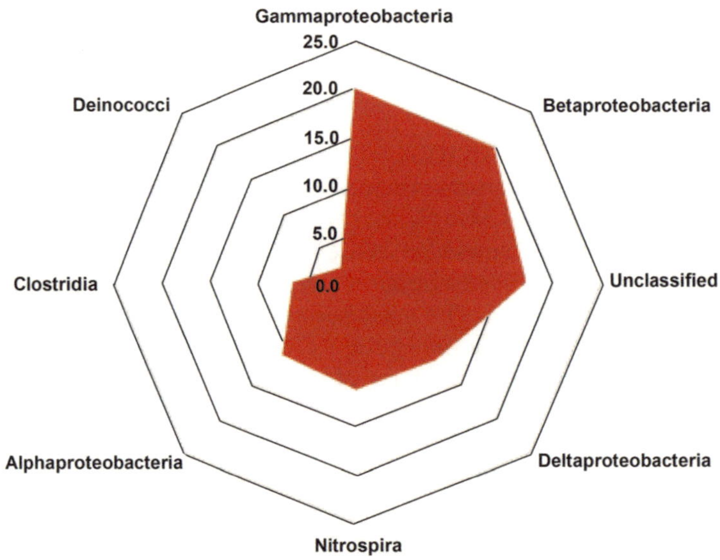

Figure 14.
Bacterial communities by class of sulfurous water from Well AC1A at Longroiva Medical Spa (from [34]).

(19.80%). Some classes were also occasionally recorded (Actinobacteria, Flavobacteriia, Nostocophycideae) in the spring samplings, with a relevant percentage of microorganisms not classified at the Class level (17.23%), with a predominance of the same in the samplings of autumn.

With regard to classification at the level of gender, there is a large variation from sample to sample. However, the genera Desulfomonile, Ectothiorodospira, and Thermodesulfovibrio are present in at least three samples [33].

According to Sá Pereira [33], in terms of species, the most representative are: Desulfomonile tiedjei, Methyloversatilis universalis, Thermodesulfovibrio aggregans, Ectothiorhodospirahalo alkaliphila, Veillonella dispar, Methylobacterium radiotolerans, Methylobacterium mesophilicum, Pseudomonas plecoglossicida, and Chondromyces pediculatus.

7. Conclusion

The Longroiva Medical Spa has been supplied by several abstractions over the last 50 years, namely, the Classical Spring and the Well TD1, which have since been deactivated, and the Well AC1A that allowed the licensing of this resource as natural mineral water and is currently still in operation.

Well AC1A is vertical, 211.7 m deep, with stainless steel tubing, cement in its annular space from the surface up to 104.3 m, and an exploration flow of 6.2 L/s as artesian (no need for a pump). It has its development in granitic rocks and captures water from a deep confined aquifer system of special and rare characteristics in the region.

The land surrounding the Well AC1A is of high to very high vulnerability, and therefore, there is a need to provide various defense mechanisms for the aquifer system; in this sense, the figure of the Protection Perimeter, namely, the Immediate Protection Zone, presents high restrictions on urban occupation and anthropic actions, in addition to the fact that a flow greater than that of natural a artesian does not withdraw from the well AC1A. There are several potential sources of pollution of the mineral aquifer, and in recent years, with the construction of the hotel in the immediate area of the protection perimeter, although with the works, there have been positive actions to decontaminate the land next to the Classical Spring, where the public washrooms used to be, anthropogenic pressure has increased along with well AC1A; this situation is not very serious immediately due to the characteristics of the aquifer, as it is confined, deep, and has a piezometric level higher than the topographic surface, but since it is of the fissured type and includes some connections to the surface materialized by several natural springs, in the medium- and long-term, there may be a risk of contamination. Thus, systematic analytical control is essential to assess the physical-chemical and microbiological stabilities of the hydromineral resource over time.

In the present work, it was possible to statistically compare two different periods: the legalization period, for 19 consecutive months (1999–2000), and the control period, over the last 20 years (2001–2020). Thus, the statistical study of the physical-chemical analyses of the water from the well AC1A during the legalization period shows great stability of most chemical elements, with RSD lower than 10%, with the exception of ammonium (NH_4^+) (10.75%). Over the last 20 years (control period), there have been some elements with RSD greater than 10%, revealing some instability, as is the case with S_T, H, K^+, NH_4^+, SO_4^{2-}, CO_3^{2-}, and $H_3SiO_4^-$. The use of multivariate analysis techniques, namely, PCA, for the two phases (1st phase includes 19 monthly analyses in a row and 2nd phase includes all 58 analyses carried out on the Well AC1A water) allowed us to understand that the characteristics of this water are determined by complex and diverse water/rock interaction phenomena, where sulfur and residence time contribute to the chemical complexity of these waters. The presence of the NH_4^+ and SO_4^{2-} species seems to be related to the sulfurous environment of these waters and not to anthropic contamination phenomena, since they are associated with the F- ion, with a possible profound influence. Regarding the comparison between the two phases, it was observed that there was a tendency toward slight differences, which can be explained by the tendency to increase the concentration of reduced sulfur species (S_T, HS^-) over time and decrease the concentration of CO_3^{2-}, a consequence of the evolution of the whole system toward stability.

Analytical results of the water from the Well AC1A were also compared with those existing in relation to the water from the other abstractions that served the Longroiva Medical Spa (Classic Spring and Well TD1), and the respective average values were projected in the Piper and Stiff diagram. From the analysis carried out, it appears that we are facing the same water and confirms the chemical stability of these waters over 50 years, although small variations in some elements have been evident, which are understood to be related to natural oscillations and even to tuning of analytical techniques.

In the case of trace elements, it was verified that the elements that are part of the typical group of these waters (B, Cs, Rb, Sr., and W) show an increasing tendency over time, with W being the element that most increased its concentration; the remaining elements do not show significant increases or trends that point to a potential situation of contamination of the mineral aquifer. The small number of analyses carried out, since the hydromineral resource is analytically controlled, does not allow the acquisition of sustained knowledge about the evolution of these elements over time in these waters.

In terms of analytical control relating to classical microbiology, carried out on water from Longroiva over time, three different periods should be considered: (a) the phase of exploring the water at the Classical Spring and at the Well TD1 (40 m deep), between 1969 and 1998; (b) the licensing phase of the resource as natural mineral water, after the construction of the Well AC1A (211 m deep), during the years 1999 and 2000; and (c) the control phase, in that same well (AC1A), over the last 20 years. From the results obtained, it can be verified that in the first phase, during the exploration of the Classical Spring and that of the Well TD1, the water systematically presented itself microbiologically inappropriate (**Table 5**); in the licensing phase, with the exploitation of water from the Well AC1A, there were no defaults for 19 consecutive months (**Table 6**); and in the control phase, over a period of 20 years, 718 analyses were carried out, with 14 analyses with values of quantities of microorganisms cultivable at 37°C, in 24 hours, greater than 5 CFU/1 mL; 4 analyses with values of microorganisms cultivable at 22°C, in 72 hours, greater than 20 CFU/1 mL; and only one analysis with total coliforms, in 2015, when the hotel was being renovated (**Table 7**). From what was mentioned above, it is understood that the potential microbiological contamination of these waters may be associated with the pressure exerted by anthropic actions along the abstractions.

Regarding the hydrogenome project, from the analysis carried out on the results obtained for the mineral waters of Longroiva, it was possible to highlight the following: (i) the waters of Longroiva can be considered chemically identical to the sulfurous, alkaline, sodium bicarbonate, and fluoridated waters, which are part of this study; however, with regard to the identified microorganisms, although it was not possible to recognize a pattern in the four samples taken, it was possible to perceive that there are differences both in typology and in the number of individuals in these waters; (ii) the diversity of microorganisms increases in the samples collected in autumn (F3-F7) compared to those collected in spring (F1 and F5), contributing to this is the increase in the percentage of microorganisms that it was not possible to classify; (iii) as these waters have great physicochemical stability due to the fact that they come from very deep aquifers, captured through wells that are also relatively deep, and are well isolated in the more superficial areas, one would expect, from the outset, more stable microbiological patterns in all samplings, and the changes that occur from sampling to sampling are surprising (**Table 8**).

Although, for now, there is no evidence of contaminating effects on this natural mineral water, it must remain under close surveillance, and all the mechanisms and good practices necessary for the preservation of the aquifer system and abstraction must be safeguarded, as provided for in the protection perimeter. The elements highlighted in the present work are guidelines for taking special care and not allowing yourself to be abused by disrespecting either what is foreseen in the Exploration Plan or the restrictions imposed by the Protection Perimeter.

Acknowledgements

FCT – Foundation for Science and Technology supported this work with Portuguese funds within the GeoBioTec Centre (Project UIDB/04035/2020).

Conflict of interest

The data sets generated during and/or analyzed during the current study are available from the corresponding author on request after the publication of work.

Author details

Pedro Jorge Coelho Ferreira[1,2*], Luis Manuel Ferreira-Gomes[2,3] and Alcino Sousa Oliveira[4]

1 Municipality of Meda, Portugal

2 Beira Interior University, Covilhã, Portugal

3 Centre GeoBioTec (UA), Aveiro, Portugal

4 Trás-os-Montes e Alto Douro University, Vila Real, Portugal

*Address all correspondence to: pedroferreirajc@gmail.com

References

[1] Ordem de cristo. Registo dos Tombos das Comendas de Castro Marim, Longroiva, Santa Maria, A Grande, Entre Outras, organizado pelo Dr. Pedro Álvares Seco. ANTT, PT/TT/OCCT/E/003/0236_m218; 1570

[2] Henriques FF. Aquilégio Medicinal. Lisboa: Lisboa Occidental; 1726. p. 348. Available from: https://purl.pt/22614/1/index.html#/5/html

[3] Almeida A, Almeida JD. Inventário Hidrológico de Portugal. Lisboa: Edição Instituto de Hidrologia de Lisboa; 1975. pp. 88-111

[4] Ferreira Gomes LM, Daniel J, Pissarra Cavaleiro VM. O recurso hidromineral das Termas de Longroiva como uma nova água mineral em classificação. In: II Seminário de Recursos Geológicos, Ambiente e Ordenamento do Território. Vila Real: UTAD; 2001. pp. CO-85-CO-97

[5] Ferreira Gomes LM, Ferreira Guedes J, Gomes da Costa TC, Coelho Ferreira PJ, Neves Trota AP. Geothermal potential of Portuguese granitic rock masses: Lessons learned from deep boreholes. Environmental. Earth Science; 2015;**73**(6):2963-2979. DOI: 10.1007/s12665-014-3605-y

[6] DR. Diário da República, IIIª Série - N.º 18 de 26 de Janeiro de 2005, para atribuição de Área de Concessão, correspondente ao número HM-53 de cadastro de designação "Longroiva"; 2005

[7] Lourenço C, Pasqual R. O Estudo Metagenómico das Águas Minerais Naturais Tendo em Vista o Reconhecimento Científico das Vocações Terapêuticas. Direção de Serviços de Recursos Hidrogeológicos e Geotérmicos da DGEG. Boletim de Minas, 53. Lisboa; 2019. pp. 39-53

[8] Directive 80/778/EEC. Council Directive of 15 July 1980 relating to the quality of water intended for human consumption. Official Journal of the European Communities; 1980. No L 229: 11-29

[9] Direção-Geral do Território. Carta Administrativa Oficial de Portugal (CAOP), shapefile, DGT, Lisboa; 2017

[10] Global country boundaries (Arcgis by ESRI). Global country boundaries from DIVA-GIS. Shapefile from: iangliangcun@gmail.com_ucsb. Available from: https://www.arcgis.com/home/item.html?id=2ca75003ef9d477fb22db19832c9554f. [Accessed on 05-05-2019]

[11] Ferreira B. O relevo de Portugal Grandes Unidades Regionais. Associação Portuguesa de Geomorfólogos. Vol. II. Coimbra: Edição Fundação para a Ciência e a Tecnologia; 2004. pp. 97-110

[12] Silva AF, Ribeiro ML. Notícia Explicativa da Folha 15-A. Vila Nova de Foz Côa. Lisboa: S.G. de Portugal; 1991. p. 52

[13] DGT. MDT50m-WMS; 2007. Available from: http://id.igeo.pt/sdg/4d0b0a9e2b4c41a886929d97b61ac459

[14] APA, IP. Atlas do Ambiente. Shapefile, APA, Lisboa; 2018

[15] Ribeiro A, Antunes MT, Ferreira MP, Rocha RB, Soares AF, Zbyszewski G, *et al*. Introduction à la géologie générale du Portugal. Serviços Geológicos de Portugal. Lisboa; 1979:114

[16] Dias R, Ribeiro A, Coke C, Pereira E, Rodrigues J, Castro P, *et al*. Evolução estrutural dos sectores setentrionais do

autóctone da Zona Centro-Ibérica. *In*: Dias R, Araújo A, Terrinha P, Kullberg JC, editors. Geologia de Portugal. Vol. 1. Escolar Editora, Lisboa; 2013. pp. 73-147

[17] SGP. Carta Geológica de Portugal, Escala 1/500000. Lisboa: Serviços Geológicos de Portugal, Lisboa; 1992

[18] LNEG. Carta Geológica de Portugal à escala 1:500000. 5ª edição da Carta Geológica de Portugal à escala 1:500000 Serviços Geológicos de Portugal; 1992. Unidade de Geologia, Hidrogeologia e Geologia Costeira. Laboratório Nacional de Energia e Geologia, I.P., Lisboa; 2019. Available from: http://geoportal.lneg.pt/arcgis/services/CGP500k/MapServer/WMSServer?request= GetCapabilities&service=WMS

[19] Meireles C, Pereira E, Ferreira N, Castro P. O Ordovícico da Serra da Marofa: novos dados litoestratigráficos e estruturais. In: VII Congresso Nacional de Geologia. Sociedade Geológica de Portugal; 2006. pp. 641-644

[20] Azevedo MR, Aguado BV. Origem e Instalação de Granitóides Variscos na Zona Centro-Ibérica. In: Dias R, Araújo A, Terrinha P, Kullberg JC, editors. Geologia de Portugal. Vol. 1. Escolar Editora; 2013. pp. 377-402

[21] Coelho Ferreira PJ. Modelação de sistemas geohidráulicos profundos associados a fraturas extensas da região da Meda. [Ph.D. thesis], Univ. da Beira Interior. Covilhã; 2022. (in press)

[22] Ferreira Gomes LM. Estudo Hidrogeológico para Enquadramento Legal das Termas de Longroiva. CMM: UBI, Relatório inédito interno; 2001. p. 48

[23] Van Duijvenbooden W, Van Waegeningh HG. Vulnerability of soil and groundwater to pollutants. *In*: Proceedings and Information No. 38 of the International Conference held in The Netherlands, *in* 1987. Delft, The Netherlands: TNO Committee on Hydrological Research; 1987

[24] Ferreira Gomes LM. Perímetro de Proteção das Termas de Longroiva. CMM, Relatório inédito interno, UBI, Covilhã; 2002. p. 52

[25] Linda A *et al*. A standardized System for Evaluation Ground-water Pollution Potential using Hydrologic Setting NWWA. OIH, USA: NTIS; 1987

[26] Coelho Ferreira PJ, Ferreira Gomes LM, Carvalho PEM, Oliveira AS. Water quality in deep and confined aquifer systems of granite rocks - the case of sulphurous water from Longroiva, Portugal. *In*: Cavaleiro et al., editors. Twin International Conferences, 2nd Civil Engineering & 5th Concrete Future, Covilhã, Portugal; 2013. pp. 26-28

[27] DR. Diário da República, 1ª Série - N.° 232 de 4 de Dezembro de 2006, fixa o perímetro de proteção para a concessão de água mineral natural denominada de "Longroiva". DRE; 2006

[28] Silva, JJRF, Almeida JD. Relatório de Análise. Água das Caldas de Longroiva (nascente sulfúrea); 1970. p. 10

[29] AMCB, DGT. Ortofotomapas à escala 1:2000. Localidade de Longroiva. Associação de Municípios da Cova da Beira e Direção-Geral do Território. Licença de utilização n.° 235/21; 2015

[30] AMCB, DGT. Cartografia digital dos Concelhos de Meda, Trancoso, Figueira de Castelo Rodrigo e Pinhel, à escala 1: 10000. Temas: eixos de vias, vias e construções. Associação de Municípios da Cova da Beira e Direção-Geral do

Território. Licença de utilização n.º 235/21; 2014

[31] Andad. Software de estatística ANDAD. Centro de Geosistemas do Instituto Superior Técnico. Versão 7.12; 2013

[32] Ferreira Gomes, LM. Estudos, Notas e Trabalhos sobre recursos hidrominerais e geotérmicos das Termas de Longroiva. Relatório Interno; UBI, Câmara Municipal de Mêda; 1999

[33] Sá Pereira. Hidrogenoma. Estudo do Microbismo Natural das Águas Minerais Naturais. Relatório Técnico de Longroiva (HM-53). Instituto Nacional de Investigação Agrária e Veterinária, I.P e Direção Geral de Energia e Geologia; 2019. p. 7

[34] DGEG. Plataforma Hidrogenoma. Direção Geral de Energia e Geologia, Lisboa; 2020. Available from: https://hidrogenoma.javali.pt/. 2020

Chapter 6

Aquifer Management in Hydrocarbon Exploitation Operations

John Fernando Escobar, Darío Naranjo-Fernández, Sergio Lopera, Orlando Mauricio Quiroz, Alonso Ocampo and German Zarate

Abstract

The exploration and production of hydrocarbons require the management of associated aquifers not only because they must be intercepted and isolated during the drilling process, but also because they can be used as reservoirs for the final disposal of connate water, or as a source to maintain the energy of reservoirs and enhanced oil recovery operations. Despite the technological advances in the management of aquifers in the hydrocarbon industry, these kinds of operations have not been free of risks and opposition. Primary oil exploitation, improved oil recovery, and disposal of remaining water operations usually involve medium and deep aquifers that have physical, chemical, and geomechanical characteristics that may differ greatly from those known in shallow aquifers. Therefore, a detailed study of the porous media as well as the contained, produced, or reinjected fluids is an obligation for the regulation compliments and the safe handling of these operations. This chapter deals with issues associated with the chemical interaction of water, the problems that may arise from an incorrect evaluation or management, and the phenomena that can show a problem or evolution in injection operations. Finally, corrective, and preventive treatments and procedures that are commonly used in the hydrocarbon industry are specified.

Keywords: oil and gas industry, shallow and deep aquifers, production waters, disposal, injection, enhanced oil recovery, hydrogeochemical and isotopic analyses, microseismicity

1. Introduction

Water requirements for oil and gas extraction are often unknown to decision-makers and public opinion. Although hydrocarbon exploitation does not involve a significant overall water consumption worldwide compared to other industries, such as agriculture or municipal needs, water demands can have serious impacts on local water resources and increase disputes between water users in high-water stress areas or during the drought seasons [1].

Oil and gas production is generally accompanied by water production with several characteristics, which affect the normal balance of the water cycle, either by water extraction or injection from and to different reservoirs. Therefore, it is required to evaluate the consequences in the water life cycle, due to water amount and source necessary for operations, water management practices, wastewater recycling, treatment, or disposal, and the impact on watersheds and the surrounding environment.

A general balance of water in industry would allow separate them into two large groups: water required for production and water resulting from extraction. These groups often have different physicochemical, geochemical, bacteriological, and isotopic features. The first one includes water with fast renewal time, low electrical conductivity, and hydrochemical and isotopic characteristics like surface water. On the other hand, the water belonging to the second group presents slow renewal times, high values of electrical conductivity, and hydrochemical facies that allow their clear differentiation from the surface water present in the region.

In most cases, water extracted together with the hydrocarbons correspond to connate water present in conventional reservoirs. They are usually a complex mixture of dissolved and dispersed organic and inorganic compounds [2]. On the other hand, production waters and return fluid consist of a formation water mixture and other fluids used during exploitation, whose composition varies spatially and temporally.

Formation waters are naturally present in all oil and gas deposits whose origin is difficult to determine in a general way. These waters are classified as endogenetic and exogenetic; in the first group, waters were originally present in the formation when the hydrocarbon accumulation occurred (connate waters), while in the second one, they infiltrated from the surface (absorbed) or penetrated from upper sediment accumulations or migrated by compaction of lower sediments (juvenile).

This work proposes alternatives to characterize these types of water, define the methodologies, and identify the most common risks, as well as highlight future trends in water resource management.

1.1 Approach

Typically, the oil reservoirs are developed initially by producing the wells by depleting the original pressure of the reservoir (*Natural Depletion*), and also taking advantage of the presence of an original gas cap and/or the solution gas liberation and expansion within the system as a consequence of the natural depletion (*Solution Gas Drive and Gas Cap Expansion*). Nevertheless, these mechanisms, referred to as primary depletion, render very low recovery factors (Usually less than 25% of the *Oil in Place* for medium and light oils, and less than 5% for *Heavy Oils*). This condition is different for the reservoirs where the oil column is connected to a very large and/or recharged aquifer[1] that can provide enough pressure support and displacement to the hydrocarbons, in which case recovery factors as high as 60–65% [3] can be obtained depending on the fluid type, continuity and homogeneity of the reservoir, and the size of the aquifer and recharge source. The gas reservoirs, on the other hand, are developed under natural depletion in most cases.

Once the primary recovery mechanism is exhausted, the reservoirs are subjected to the injection of water and/or gas for pressure maintenance and

[1] In reservoir engineering terminology a large aquifer refers to aquifer whose volume is a thousand times o higher than the hydrocarbon volume. While recharged aquifers which are connected to surface water bodies.

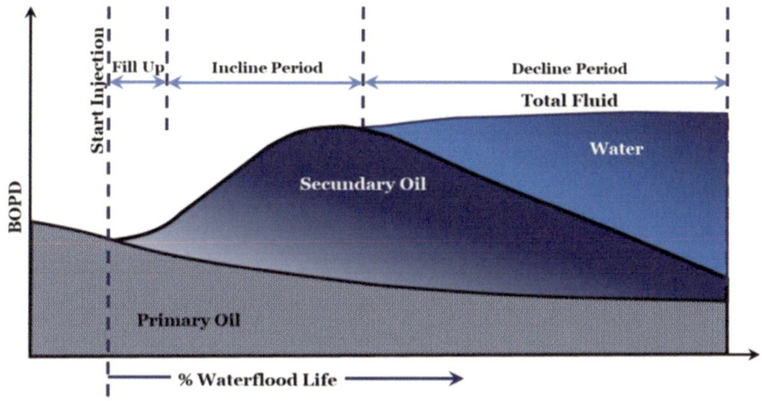

Figure 1.
Typical performance of a successful waterflooding project [5].

sometimes for displacement processes. This stage is referred to as secondary recovery and can provide additional recoveries factors between 10 and 30% over the primary recovery. Finally, some reservoirs are subjected to enhanced oil recovery processes that can be of different types: Chemical processes based on the injection of water plus specific chemicals to improve the sweep and displacement efficiency of the project, and thermal methods such as cyclic and continuous steam injection and in-situ combustion based on air injection. Other enhanced oil recovery methods include CO_2 and enriched hydrocarbon gas for injection for miscible displacement.

So far, waterflooding is the most widespread oil recovery mechanism implemented worldwide in the oil industry, due to its technical simplicity, cost-effectiveness, and the availability of the resource in most of the provinces where oil projects are developed. Most of the mature oil production development is accompanied by significant volumes of water. This condition makes the reinjection of these production waters for pressure support and displacement of hydrocarbons within the same producing horizons one of the best options, not only from the environmental side but also from the technical and economical optimization side.

Now, a water injection project for pressure maintenance and oil sweep improvement involves both surface and subsurface matters. The subsurface domain includes reservoir engineering, geology, and geophysics as well as the production technology; while the surface issues include, among other requirements, the water injection source, and the water treatment and injection infrastructure [4].

Figure 1 illustrates the performance of a successful waterflooding project. During the primary recovery stage, the oil rate declines continuously due to the pressure reduction with time, due to the fluids offtake. Once injection starts, the injected water starts restoring the pressure of the reservoir and getting the liberated gas back in solution; during this period, referred to as *Fill Up*, no incremental oil production is observed.

After the *Fill Up* period is completed, the displacement process starts acting and the effect on the well producers is observed via a ramp-up in production, whose slope and summit depend upon the homogeneity of the reservoir, the mobility ratio (M), and the amount of water injected. Finally, when the water breaks through the oil production wells, the *Water/Oil Ratio* starts climbing up and the oil rates decline until the end of the project, when the oil rate gets to its economical limit.

Therefore, it is of vital importance to have a first conceptual approach to the characteristics of the medium and the fluids contained in order to later characterize the volumes of water, its origin, its physicochemical quality, isotopic composition, risks during handling in production or injection operations and treatment alternatives according to their composition and volume. These methodologies are commonly used in the industry, but in some cases, they are not rigorously applied or with the participation of hydrogeologists with experience in medium and deep aquifers.

1.2 Methodology

To assess and manage hydrological resources in a region, the first step is to build a conceptual hydrogeological model (CHM) that includes geological, hydrodynamic, hydraulic, geochemical, and isotopic information. In this model, analyses of the fluids, inferred flow paths, and cross-connections should be included, based on a holistic understanding of the physical and chemical framework of the geological and hydrological environment. CHM should also establish the hydrochemical characteristics, isotopic signatures, and residence times of different aquifers including, besides, the possible sources of current system recharging. This will allow the establishment of the interaction between the different water reservoirs in areas where the projects are located.

Finally, CHM should serve as the basis for mathematical flow and transport models, which allow the quantification of the water volumes involved in the processes and contribute to identifying relationships not considered that may occur between the different fluid reservoirs. However, it should be clear that these mathematical representations are only expressions of current knowledge of the system and should be continuously reviewed and improved.

Beyond the initial conceptual model, the phenomena and processes that can occur at the formation level are dominated by the characteristics of the water produced or injected and the characteristics and affinity of the producing or receiving environment that, in most cases, have the conditions of a confined aquifer. A large part of the operations carried out come from hydrocarbon reservoir engineering, which has traditionally considered water as a by-product or waste; but that more recently has assumed these waters as a source of water supply or, at least, it has advanced in water consumption reduction and minimizing environmental impacts. Regardless of the above, in the management of water associated with the production of hydrocarbons, it is mandatory to characterize the volumes to be handled, physicochemical and isotopic characteristics, and consequently select the method of treatment or disposal of these fluids, as will be explained below.

2. Aquifer and water management associated with hydrocarbon operations

Interaction of hydrocarbon exploration and production operations with water present in several subsoil formations begins during the drilling process of the surface section of the exploratory or development wells. Operations interaction tends to be controlled and, in fact, bypassed using appropriate drilling muds, clay mud cakes (which seal the flow from the well to the formation and vice versa); cementation, and steel casings (2 or 3 concentric strings). Casing and cementation play a role of double protection (protecting formation from well fluids invasion and the well from formation fluid irruption).

However, it is the operations related to the exploitation of hydrocarbons that more intensely compromise the management of large volumes of water and the formations that contain them, either because they are used as sinks for the disposal of effluents of water produced with hydrocarbons (e.g. Disposal), or because they are a source of water for enhanced recovery processes (e.g. EOR-Reuse) or because they are reinjected into the oil-bearing formations to maintain reservoir pressure (e.g. Reuse-Reinjection).

2.1 Example of magnitudes and volumes in the management of groundwater in the hydrocarbons industry

Colombia has an average daily oil production of 750,000 barrels and associated water production of close to 13 million barrels. This 1:17 ratio tends to be stable with small variations, mainly due to certain operations that may be temporarily stalled due to reconditioning works or excessive increases in water cuts in new fields.

This water production is monitored and controlled by the National Hydrocarbons Agency (ANH), through the mandatory production daily report provided by the Oil and Gas operating companies. This way, the efficiency of the operation can be controlled, as well as the compliance with the commitments contained in the environmental license, which is linked mainly to the proper disposal of these waters and the protection of aquifers. **Figure 2** shows the historical and synchronous monitoring of these volumes through the ANH Control Center, which makes it possible to determine not only the volumes by country and region, but also by field, formation, contract, and even by well.

Petroleum companies have shown adequate behavior in the management of groundwater and actively participate in the reporting of information on the quality and quantity of water used and final disposal. This behavior responds not only to compliance with the regulations but is also part of its continuous environmental improvement programs. For example, the national hydrocarbons company (ECOPETROL), which for the date exemplified in **Figure 3** produces 60% of oil and 55% of the associated water in the country, develops continuous projects that allow increasing the reuse of production waters, thus reducing the collection of surface sources and shallow aquifers, while reducing the volumes to be injected into final disposal [7].

Under the guidance of the ministries of Mines and Energy and the Ministry of Environment and Sustainable Development, initiatives are oriented so that hydrocarbon projects that can generate large impacts on the aquifer have the pertinent information on ecosystems, health, hydrology, and hydrogeology through free access WEB portals for all audiences [8]. These measures are complemented with regulatory solutions that allow government entities with greater technical strength and thematic knowledge to regulate the use (as a source or sink) of medium and deep aquifers in the hydrocarbons industry, establishing a vertical border, to determine the regulation power, to the so-called regional hydrocarbon seal, understood them as the layer of clay that regionally seals the migration of hydrocarbons to the surface [9]. **Figure 4** illustrates the

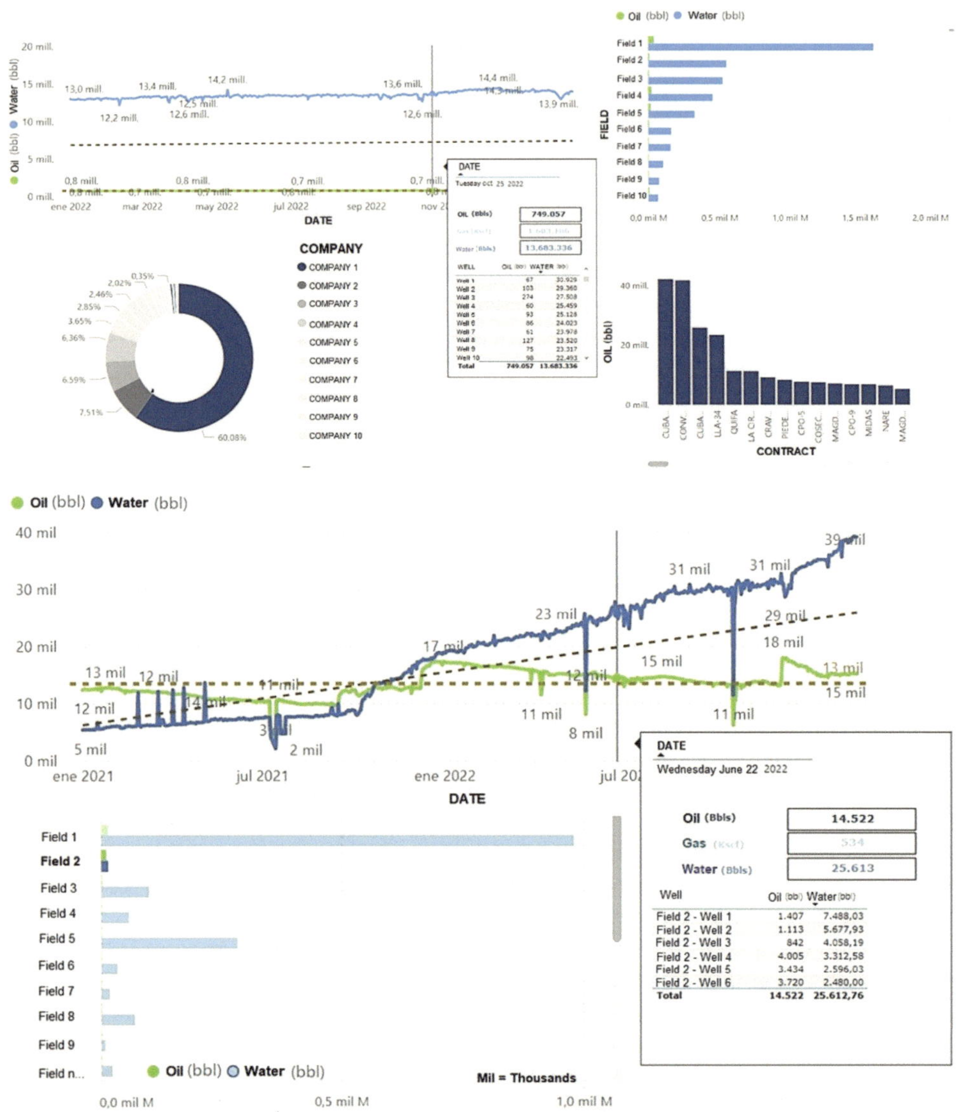

Figure 2.
(Above) Typical configuration of one dashboard for monitoring the production of water associated with oil and (Below) example of the increase in water production (x8) in a specific field (Field 2) in a period of 24 months [6] (To preserve the confidentiality of the information and a better understanding, the graphs were translated and simplified using a graphic editor.).

distribution of possible contacts of the wellbore with interest formation (left), as well as components and possible geomechanical issues that could occur in a disposal or EOR project (right).

Notwithstanding the foregoing, beyond the regulations, the contingencies that may arise in these operations are largely related to the nature of the water that is managed, the physicochemical processes that may occur at the formation level, the affectations that may suffer the porous media, and the technical variables of the reinjection operation itself, as will be seen later.

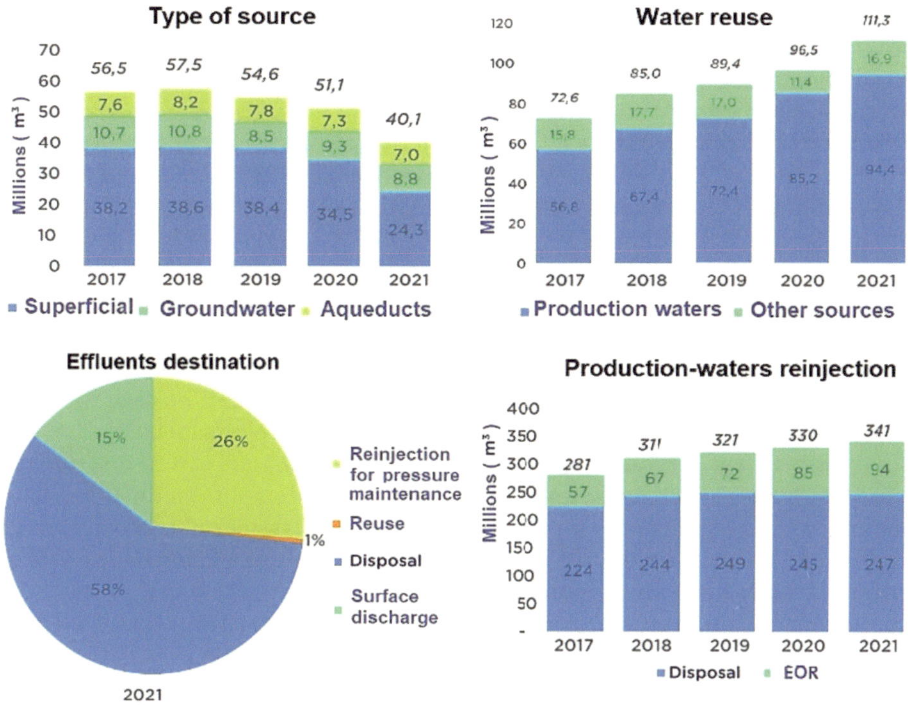

Figure 3.
Achievements reported by ECOPETROL in the period 2017–2021 [7].

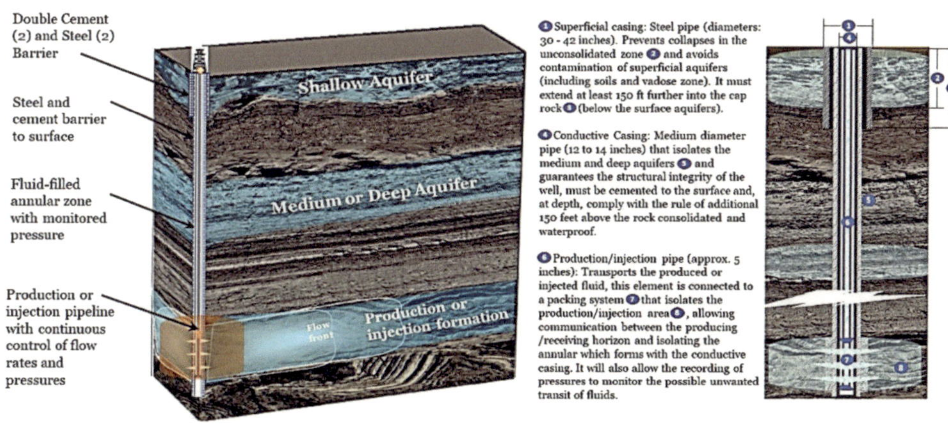

Figure 4.
(Left) distribution of geomechanical elements in the wellbore, and (right) characteristics and function of this component Vis a Vis the saturated formations that can be intercepted [Created by authors].

2.2 Typical composition of formation waters

Formation waters composition depends on factors such as endogenous waters composition in sedimentary rocks linked to depositional environments; subsequent changes due to rock-hydrocarbon-water interaction amid sediment compaction;

changes due to rock-hydrocarbon-water interaction during migration (if it occurs) and changes due to mixing with exogenous water, including younger water such as meteoric water. Among typical components of these waters are inorganic components that are present in significant quantities ranging from the order of thousands of mg/L for sodium (Na^+) and chloride (Cl^-) ions going through the order of mg/L of ions such as calcium (Ca^{2+}), magnesium (Mg^{2+}), bromide (Br^-), sulfate (SO_4^{2-}), sulfide (S^{2-}), potassium (K^+), strontium (Sr^{2+}), carbonate (CO_3^{2-}), aluminum (Al^{3+}), iron (Fe^{2+}), barium (Ba^{2+}), lithium (Li^+), ammonia (NH^{4+}) and borate (BO_3^{3-}), up to the order of thousandths of mg/L for ions such as manganese (Mn^{2+}), silicate (SiO_3^{2-}), iodide (I^-), chromium (Cr^{3+}), copper (Cu^{2+}), nickel (Ni^{2+}), lead (Pb^{2+}), fluoride (F^-), phosphate (PO_4^{3-}) and arsenate (AsO_4^{3-}). Knowledge of this composition is vital to identify and evaluate mud formulation during drilling operations and potential treatments, either for disposal or reuse. This information can also be used for well-log analysis, environmental impact assessment, and geochemical exploration.

On the other hand, organic components of formation waters can be dissolved and dispersed; they are composed mainly of aliphatic and naphthenic acids with anions and, in less concentrations, amino-acids and aliphatic, cyclic, and aromatic hydrocarbons. Determination of dissolved organic compounds in formation waters is important to study hydrocarbon-associated phenomena, such as its origin and migration or its disintegration and degradation. Further, large amounts of dissolved gases, mainly hydrocarbons are contained in formation waters; however, inorganic gases such as carbon dioxide (CO_2) and hydrogen sulfide (H_2S), associated with corrosion processes, and nitrogen (N_2) are frequently present [10].

Other physicochemical properties like the pH and the redox potential are very important. Knowledge of pH and redox potential allows the evaluation of the possible formation of scales due to its influence on the solubility of different elements and components, and the corrosion tendency of water mainly associated with hydrogen sulfide.

2.3 Typical composition production waters

Production waters is the major by-product waste associated with hydrocarbon production and usually increases as reservoir declines in hydrocarbon production. Components of produced water can be grouped into the following categories: suspended and dissolved solids, anions, metals, radionuclides, and organic compounds. Some inorganic compounds are present in production waters in substantially higher concentrations than marine waters. These compounds include sodium, chloride, barium, iron, manganese, mercury, and zinc. High concentrations of multivalent species such as iron and manganese may also be present due to low redox potential values (anaerobic conditions) in Formation waters, which favor the predominance of reduced soluble species of these ions. When extracting production waters, these reduced ions will readily precipitate as ferric hydroxide [$Fe(OH)_3$] or manganese oxide (MnO_2) due to chemical oxidation caused by contact with atmospheric oxygen. Other elements, such as zinc and probably lead, could derive in part from galvanized steel structures in contact with production waters. Less common may be radioisotopes from naturally occurring radioactive material (NORM), contained mainly in formation waters, or injected marine water. The most abundant radioisotopes are radium-226 (^{226}Ra) and radium-228 (^{228}Ra), derived from the natural decay of uranium-238 ($_{238}U$) and thorium-232 (^{232}Th) associated with certain reservoir rocks and clays.

Organic components in production waters can appear as dissolved or dispersed [11]. Among the dissolved organic components are saturated petroleum hydrocarbons, aliphatic or aromatic, but due to the higher solubility of aromatic compared to aliphatic of the same molecular mass, dissolved aromatic hydrocarbons predominate, mainly Benzene, Toluene, Ethylbenzene, and Xylene (BTEX), and Polycyclic Aromatic Hydrocarbons (PAHs) of lower molecular mass such as naphthalene and phenanthrene (two to three benzene rings). Other dissolved organic compounds are phenols, compounds of interest due to the toxicity of some alkylphenols as endocrine disruptors. Among the dispersed organic compounds, due to their low solubility, PAHs with a higher molecular mass (more than three benzene rings) are usually found, together with saturated aliphatic hydrocarbons.

On the other hand, aliphatic and naphthenic carboxylic acids, produced by hydrated pyrolysis or by anaerobic biodegradation of saturated aliphatic and aromatic hydrocarbons, respectively, can occur as dissolved or dispersed organic components, depending on their molecular mass (solubility). Some authors [12, 13] have proposed different intervals of inorganic and organic components in production waters, however, detailing them would exceed the scope of this chapter.

Produced water may also contain other organic components that are used for three purposes: 1) to solve specific well problems, 2) to treat produced water intended for injection or discharge, or 3) to improve hydrocarbon recovery and pumping. In the first case, organic compounds are used to: protect the system against corrosion (oxygen scavengers, corrosion inhibitors, micro-biocides), prevent scale formation (scale inhibitors), dissolve paraffin deposits (solvents) and prevent the formation of methane hydrates in the production of gas (antifreeze); in the second case it is used to remove colloids (coagulants and flocculants) and facilitate the gas-crude-water separation (solvent emulsions, defoamers); while in the third case, polymers and surfactants are used [14, 15]. The concentration of these organic compounds in the production waters will be determined mainly by the doses used, the ratio division constant between the organic (hydrocarbon) and inorganic (water) phases, and the temperature.

3. Discussion

Nowadays, most of the world's oil fields produce with *Water–Oil Ratios* (WOR) greater than 90%; for this reason, the reinjection of production waters into oil-bearing formations is the most recommended way of disposing of these enormous volumes of water.

Other disposal options are the final confinement in other aquifer formations (deep or shallow); the reuse for production operations or surface disposal through total or partial evaporation (using lagoons or forced dispersion systems); agricultural irrigation, livestock, industry, civil works, and watering roads.

For reuse purposes, the following factors should be considered: 1) production waters generation rate compared to the water demand for reuse, 2) production waters quality and treatment requirements for reuse, 3) costs of produced water reuse compared to discharge alternatives, 4) availability of infrastructure and treatment alternatives, and 5) regulatory considerations [16]. Different reuse alternatives can be considered as potential options, implementing different levels of treatment depending on the quality of the wastewater and the desired reuse requirements.

On the other hand, the disposal in aquifers should receive special attention, since it can cause groundwater contamination if the injected fluid migrates, accidentally or deliberately, into an exploitable aquifer. This could be due to poor well design or construction, deteriorated pipelines, or a poor understanding of hydrogeology. In the last case, migration of reinjection water through cracks in confinement zones under induced pressures is common [17].

Figure 5 outlines the possible management alternatives (disposal or reuse) of production waters, classifying them into raw and treated production waters. For conventional reservoirs, one of the main alternatives for the reuse of produced water is the maintenance of pressure in the reservoir. Given the growing interest in the recovery of heavy oil and tar sands from oil-bearing formations, production waters can be heated to be injected as steam [15, 18]. The disposal alternative is not applicable in unconventional reservoirs because the low porosity of the formation offers a minimum pore volume to be saturated with water [19].

For unconventional reservoirs, production water can be reused in other hydraulic fracturing operations. In these reservoirs, the main sources for the formulation of the hydraulic stimulation fluid are superficial freshwater bodies and aquifers [20, 21]. For this reason, the reuse of production waters for hydraulic stimulation is considered

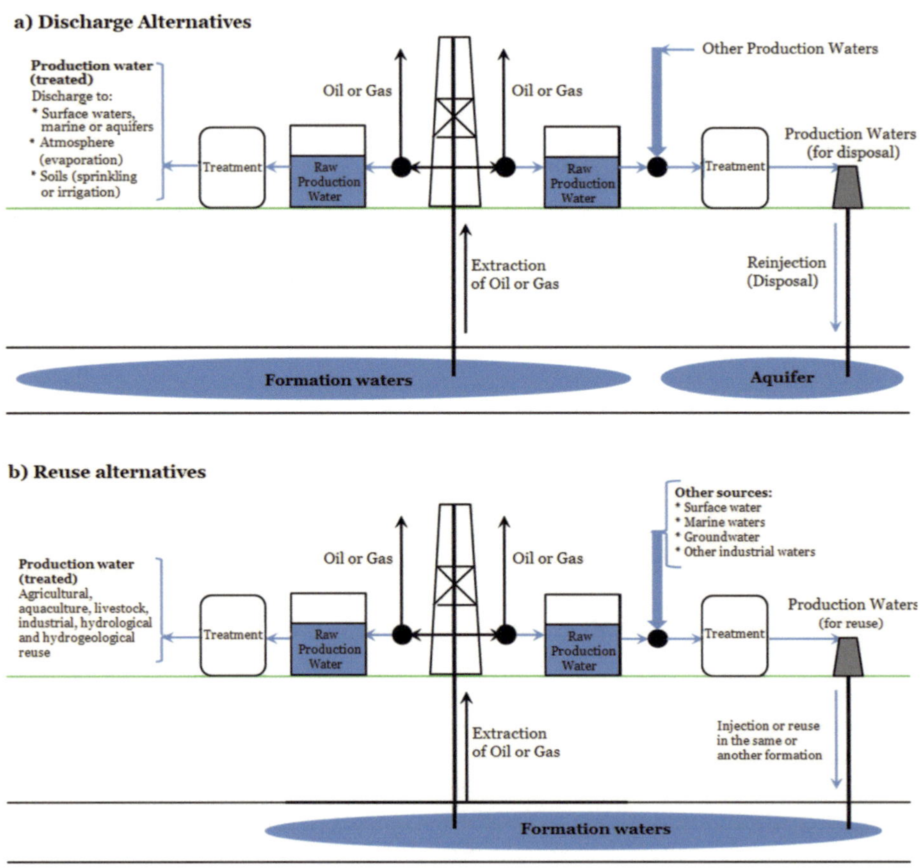

Figure 5.
Origin and possible destinations of production waters for: a) discharge (as disposal) or b) reuse (mainly in pressure maintenance operations). [Created by authors].

an alternative that allows for reducing the pressure on the water resource [14, 22]. Additionally, this practice can result in economic benefits by reducing the costs associated with the acquisition and transport of water for stimulation, and transport and reinjection in injection wells [20, 23].

In pressure maintenance, enhanced recovery, or in hydraulic fracturing operations, the production waters without treatment (raw water) needs to achieve the physicochemical characteristics of compatibility with formation waters, therefore it is possible to mix them with surface, marine, underground or production waters from other wells. However, in the event of disposal, and to reach that compatibility, the only recommended mixing is with production waters from other wells.

In spite of the above considerations, there always exists the possibility of affecting negatively the aquifers surrounding the operations of reuse and disposal of hydrocarbon-associated waters. It is important to clarify that the disposal of water is done in specific formations, i.e. geological strata which have enough porosity and permeability to produce or receive fluids through wells drilled across them. The system is comprised of wells in a spatial distribution, production facilities, and other surface flows and components that form what is called an *Oil or Gas Field*.

Although the number of wells drilled in some formations can be counted by dozens or thousands, not all of them are dedicated to water injection; nevertheless, their number and integrity status could be considered as an indirect vulnerability indicator. Despite the magnitude of the number of wells drilled worldwide, just a few cases of massive or regional affectation of aquifers are reported in the literature. The reported events are mainly associated with fluids migration to shallow aquifers, loss of integrity due to seismic events, and geomechanical or structural failures in some fields.

These unwanted events can be attributed to poor or sub-standard practices in some hydrocarbon exploitation operations, which lead to loss of integrity and structural or geomechanical failures; causing the unwanted arrival of fluids to shallow aquifers used for human activities, and subsidence with the affectation of surface infrastructure. A great deal of these unwanted consequences happened because of seismic events from a different origin, as discussed below.

3.1 Issues associated to water injection for disposal

The injection in deep aquifers is a widely spread practice for the disposal of waters associated with oil or gas production. It is accepted in most of the countries where hydrocarbon exploitations take place if the regulatory restrictions are fulfilled. It can be considered a safe practice if all the technical and environmental measures are considered in the planning and execution of the disposal operations.

In this scenario, the aquifer may be confined or could be connected to recharge zones on the surface. In the cases of confined aquifers, the injection volumes are restricted due to the extremely low compressibility of the water ($3x10^{-6}$/psi), which is in the same order as the rock compressibility [24]. In these cases, extremely large aquifers are required for a commercial-scale disposal project. In most of the cases, where this disposal process is implemented massively, the aquifers have some sort of communication with surface water sources but with renewal times long enough to avoid the injected water coming to the surface. The main challenges faced by these projects include:

- Lack of enough injectivity in the target reservoirs where the water disposal will take place due to low permeability rock or the effect of formation damage mechanisms.

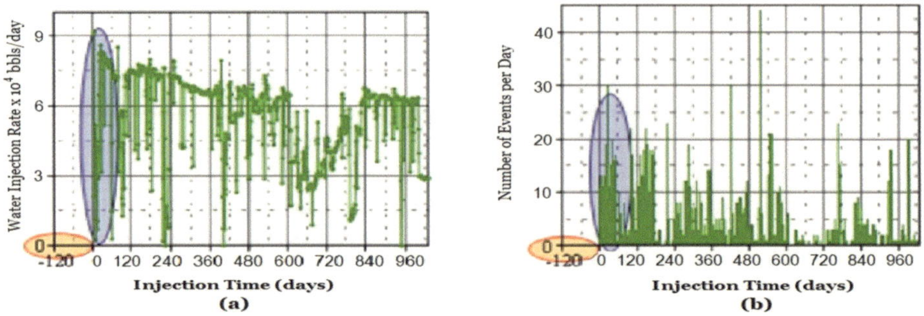

Figure 6.
Correlation between water rate injection and number of microseismicity events for the Mirador formation in the Cusiana field in Colombia [25].

- Risk of contaminating shallow aquifers that could be used for human activities, due to integrity failures in wells completion.
- Chemical shock fronts are created in the cases of significant differences in salinity between the injection and in-situ waters in the target injection reservoir.
- Generation of excessive or out-of-norm micro-seismicity due to high well-head injection pressures over injection. **Figure 6** below is an example of the correlation between water rate injection and number and microseismicity events.

To overcome or mitigate the above issues, careful analysis and evaluation of the water quality and its compatibility with the reservoir fluid should be done. These studies should include physicochemical and isotopic analyses, and local and regional hydrogeology studies to understand the size and lateral extension of the target reservoirs, and its possible connection with surrounding aquifers and/or surface water bodies.

To avoid excessive induced seismicity, both the volumes and the wellhead pressures should be controlled and closely monitored, in order to prevent injecting the water very closed to or above the fracture pressure of the reservoir. For projects that inject large volumes of water, it is advisable to install local microseismicity networks to monitor this activity.

3.2 Issues associated to water injection for reuse and EOR operations

Operations that require the use of large volumes of water combined with chemical mixtures, recirculation, injection, pressure management, and effluent treatment, among many others, are not exempt from contingencies that may influence the entire project, including the receiving or producing formations, the mechanical assemblies or even the surrounding environment at different scales and compartments.

Usually, it is common to find problems associated with the loss of injectivity, corrosion of mechanical elements, induced seismicity, and regional management of aquifers, including possible alterations of neighboring formations, either due to a possible intercommunication between different aquifers or between the oil-bearing formations and aquifers or due to aquifers use as a water source for EOR processes. Also, the channeling of the fluids due to reservoir heterogeneities, unwanted fracturing of the reservoir, and loss of containment capacity of the seal rock due to high injection pressures; may affect the technical, environmental, or financial viability of these projects.

3.2.1 Channeling and induced seismicity

One of the issues of special interest is the channeling of the fluids due to reservoir heterogeneities rendering early and fast water breakthrough at the producing wells, and generation of microseismicity events higher than allowed due to high well-head injection pressures and/or over-injection., **Figure 7**.

Higher energy-induced seismicity, derived from the injection of large volumes of waste fluids, is generated by two main causes: 1) an increase in pore fluid pressure, and 2) a change in stress state that can cause the reactivation of existing faults or fractures [26, 27]. Recent numerical models suggest that fluids travel up to hundreds of meters, while pore pressure extends to distances on the order of kilometers [28].

The occurrence of induced seismic events during residual water injection operations seems to be an inevitable process. Therefore, efforts to reduce the magnitudes of the most significant events should be aimed at replacing them with a cloud of many smaller events with equivalent total energy [29].

3.2.2 Loss of injectivity of the formation

The waters that are injected into the subsoil can interact, at the pore level, with the receiving formation, blocking the flow or reducing it. The most common phenomenon, which tends to occur at this scale, is plugging by particles penetrating into the rock, caused mainly by (a) External particles carried into the formation by injected fluids, including drilling muds, used in the completion and well repair and recovery processes; (b) Particle mobilization in situ due to drag forces and rock-fluid interactions; and (c) Appearance of particles in the formation by chemical reactions that cause organic and inorganic precipitations.

From the point of view of water chemistry, these phenomena can be grouped into:

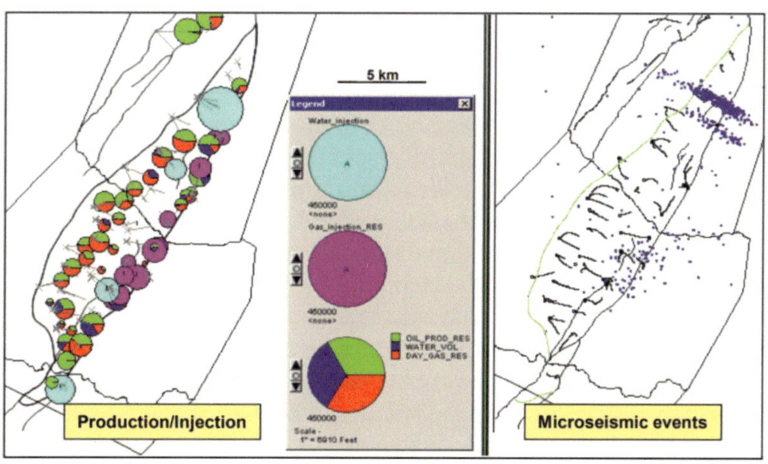

Figure 7.
Correlation between water injection wells and rates to the microseismicity and density of events in the Cusiana field in Colombia. There is also a strong alignment of the events with the direction of the maximum horizontal stress (NW-SE) [25].

Formation of precipitates (flakes and scales). Scale can reduce the permeability of producing formations, clog hydrocarbon-producing wells (in primary or secondary recovery), or injection wells. These scales are produced by some of the most common ions in formation waters that generate precipitates in case the reinjection and formation waters are not compatible. The most common scales are calcium sulfate (anhydrite), calcium carbonate (calcite), strontium sulfate (celestite), barium sulfate (barite), and iron hydroxides. The appearance of these scales depends on factors such as: pH, temperature, type, and concentration of ions, the composition of the formations, and partial pressure of the gases (for example, partial pressure of carbon dioxide in the formation of calcite). A special case is dolomite, which can be produced by the reaction of magnesium present in the formation of water in contact with calcite or by the interaction between magnesium sulfate and calcite.

On the other hand, the release of divalent ions such as calcium, magnesium, strontium, and barium into solution can cause a decrease in formation permeability due to the precipitation of chemical agents present in the "enhanced" waters used in some reinjection processes (such as polymers and surfactants). The formation of these precipitates can be controlled using inhibitors.

Dispersion and hydration of clays. The clays present in the formations, especially smectites, and illites, can decrease the permeability due to their dispersion (migration) or hydration (swelling) caused by the disposition of non-compatible waters with high pH, due to the disposition of non-compatible waters with high pH, this causes an increase in the negative surface charge and the consequent increase in the electrostatic repulsion forces. Thus, for example, the exposure of a consolidated clayey sandstone to a fluid with a high pH can reduce its permeability one hundred times in a short period of time [30]. This effect can be minimized by making the pH compatible between the reinjection and formation waters.

For its part, hydration produces a decrease in permeability due to swelling caused by the adsorption of water on the surfaces and inside the clayey structures. This originates from the disposal of waters that are not compatible with low salinities. In the case of smectites, for example, water can penetrate between the layers that compose it and cause the sheets to separate and swell. This effect can be minimized by increasing the salinity of the reinjection water (usually with sodium chloride) or by preparing the reinjection water with mixtures of water from different formations.

Anhydrite hydration. Anhydrite can be present in sandstone formations as a cementing agent between the grains of sand; if this anhydrite encounters water, it can be transformed into gypsum. In contact with high salinity solutions (such as many formation waters), the anhydrite remains unchanged, but if the reinjection waters gradually decrease the salinity of the formation, the anhydrite is hydrated, transforms into gypsum and the mineral volume increases to 1.5 times.

Presence of microorganisms. Reinjection waters may contain heterotrophic and autotrophic bacteria; Examples of the latter are the genera *Crenothrix, Gallionella, and Sphaerotilus* [10], which use the oxidation of ferrous iron, to ferric iron with oxygen, as an energy source and the carbon dioxide as a carbon source.

Among the heterotrophic bacteria are bacteria of the genera Aerobacter, Bacillus, Escherichia, Flavobacterium, and Pseudomonas, which use oxygen oxidation of various organic compounds to carbon dioxide as a source of carbon and energy. Additionally, other microorganisms such as copepods, diatoms, and dinoflagellates are present in reinjection waters when seawater is used for pressure maintenance or enhanced recovery. All this biomass can generate pore blockage either due to its detrital nature and/or due to the production of exopolymers as a consequence of its biological activity.

Fine particle migration. Because of the processes that occur in the porous media, one of the phenomena that require the greatest attention in modern fluid injection operations for EOR or disposal is triggered, this phenomenon is known as "particle migration", in Ref. to the movement of particles between 0.5 and 40 microns (range traditionally reported by specialized literature). The flow of these "fine" elements includes a diverse group of mineralogies that must be determined with specialized laboratory core analysis [31].

Particle flow processes can be classified into two groups: internal and external processes. The external correspond to those that occur on the face of the formation; while the internal occur in the porous medium; In turn, each group can be divided into three types of processes [32]: (i) those that occur on the pore surface (deposition and removal); (ii) in the throat of the pore (plugging and unblocking) and/or (iii) that affect the total volume of the pore, such as formation and disappearance in situ of crusts; migration, appearance and disintegration of particles, with or without chemical reactions; release of fine particles by chemical dissolution of the cement, coagulation or disintegration, among many others.

The most common phenomenon that tends to occur at this scale is plugging by particles penetrating the rock, caused mainly by external particles carried into the formation by injected fluids. As fine particles move along tortuous channels, they are eventually captured, retained, and deposited in the porous matrix, resulting in changes in porosity and permeability. **Figure 8** illustrates some of these processes.

These particles can be stabilized by applying chemicals that modify the forces that act between the particles and the rock. Particle stabilization studies are based on the design and implementation of displacement tests to evaluate the effectiveness of certain treatments to control the mobilization, dispersion, and generation of fine particles in consolidated cores and sand packs.

Corrosion of conduction tubes, liners, and lines. The wear and eventual rupture of the pipes due to corrosion can generate an increase in injectivity, not related to formation, but to the leakage of reinjection water towards layers of greater permeability such as superficial aquifers.

Electrochemical corrosion is a common phenomenon in injection and producer wells. This corrosion is generated by the presence of dissolved gases in the formation waters such as hydrogen sulfide (H_2S), carbon dioxide (CO_2), and oxygen (O_2), the latter present in the reinjection waters due to the contact of the production waters with the atmosphere during the crude oil separation processes. These gases are electron acceptors that solubilize the elemental iron (Fe) present in the steel, acting as an electron donor, and oxidizing it to ferrous iron (Fe^{2+}). Among the dissolved corrosive

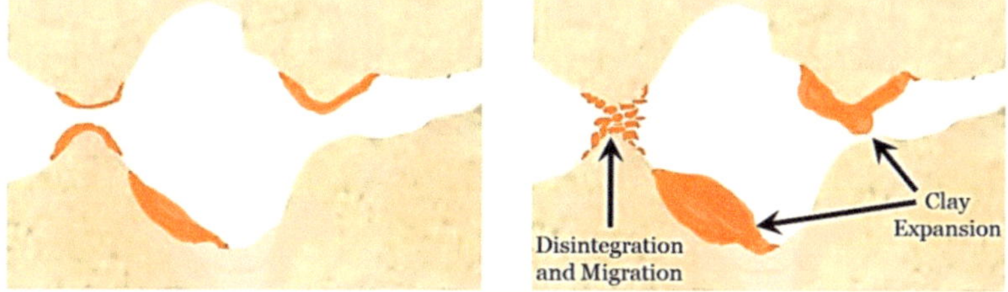

Figure 8.
Illustration of some processes that can take place at the pore level and have a direct influence on injectivity. [Created by authors].

gases, oxygen has the worst consequences since concentrations as low as 1 mg/L can cause severe corrosion [10].

On the other hand, the carbon dioxide in the solution increases as the pH decreases, due to the displacement of the carbonate system, increasing the corrosivity of the water. It must also be taken into account that with the increase in pH, the formation of calcite incrustations is favored due to the displacement of the equilibrium towards the formation of carbonate ions, and ferrous carbonate may appear, which is an incrustation that, on the one hand, can obstruct the formation, but on the other, in conditions of rapid and uniform nucleation of the crystals on the metallic surface, can constitute an additional protective layer against corrosion in pipes and conduction lines.

Solution salinity (dissolved solids concentration) has adverse effects on corrosion. On the one hand, because water acts as an electrolyte (electron conductor), corrosion is favored by increased salinity. However, the increase in salinity decreases the concentration of dissolved gases, in turn reducing the corrosion caused by them. Other factors of interest that act on this phenomenon are temperature and pressure, since the increase in temperature decreases the concentration of dissolved gases, while the increase in pressure favors it.

Microbiologically mediated corrosion can also occur. For example, the presence of hydrogen sulfide in formation and reinjection waters is of biogenic origin, produced by the metabolism of sulfate-reducing bacteria (heterotrophic anaerobic) such as *Desulfovibrio desulfuricans*, which use various organic compounds as carbon and energy sources. Because of them, ferrous sulfide (triolite) can be produced, which is an incrustation that can cause the plugging of the formation. However, under proper pH conditions, it can be deposited as a protective layer on the metal surface. Another scale that can form is ferric sulfide which is produced by the reaction of ferric oxide with hydrogen sulfide.

Finally, changes in temperature, such as those that can occur in injected formation, significantly influence all equilibria, whether chemical, precipitation/dissolution, or oxidation/reduction; or physical equilibria such as hydration with anhydrite.

4. Conclusions and recommendations

Management alternatives for water produced in oil projects include dumping or reuse. In this frame, the predominant alternatives are disposal in aquifers through injection wells (reinjection water) and reuse for improved recovery in oil-bearing reservoirs, therefore, it is essential to have a list of physicochemical parameters that allow the characterization of formation and reinjection water (**Table 1**).

Likewise, it is advisable to have a physicochemical characterization of the different aquifers present prior to the start of oil production projects.

The selection of the treatment alternative for production waters will depend, among other factors, on its physicochemical characteristics and the volume of discharge or reuse, these aspects will make it possible to determine the levels of contaminant removal, which must be within the limits established in the local environmental regulations.

The physicochemical characterization of the waters intervenes in the oil project and the geological knowledge of the formation provides indications about different

	Parameter	Unit	Problem
Generals	pH	—	Scale formation, electrochemical corrosion, clay dispersion, solution compatibility.
	Temperature	°C	
	Relative density	—	Compatibility of solutions.
	Redox potential -Eh- Total Suspended Solids (TSS)	mV	Scale formation, electrochemical corrosion
	Conductivity	μS/cm	Scale formation, electrochemical corrosion, hydration of clays and anhydrites.compatibility of solutions.
	Salinity	mg/L	
	Total Dissolved Solids (TDS)	mg/L	
	Total Suspended Solids (TSS)	mg/L	Migration of fine particles
Organic matter	Total Organic Carbon (TOC)	mg/L	Growth of aerobic and anaerobic heterotrophic microorganisms
	Chemical Oxygen Demand (COD)	mgO2/L	
	Total Hydrocarbons (HTP)	mg/L	
	Greases and oils	mg/L	Formation plugging.
Cations	Aluminum (Al^{3+})	mg/L	Scale formation
	Calcium (Ca^{2+})	mg/L	
	Barium (Ba^{2+})	mg/L	
	Strontium (Sr^{2+})	mg/L	
	Magnesium (Mg^{2+})	mg/L	
	Manganese (Mn)	mg/L	
	Iron (Fe)	mg/L	Scale formation, electrochemical corrosion, aerobic autotrophic microorganisms, microbiologically mediated corrosion.
Anions	Bicarbonate (HCO_3^-)	mg/L	Scale formation
	Carbonate (CO_3^-)	mg/L	
	Sulfate (SO_4^{2-})	mg/L	Anaerobic heterotrophic microorganisms, microbiologically mediated corrosion.
	Sulfite (S^{2-})	mg/L	Scale formation, microbiologically mediated corrosion.
Other Ions	Zinc (Zn^{2+})	mg/L	Scale formation
	Copper (Cu^{2+})	mg/L	
	Chromium (Cr)	mg/L	
	Nickel (Ni^{2+})	mg/L	

	Parameter	Unit	Problem
Nutrients	Nitrate ($N\text{-}NO_3^-$)	mg/L	Growth of autotrophic and heterotrophic microorganisms
	Ammonia nitrogen($N\text{-}NH_3$)	mg/L	
	Orthophosphates ($P\text{-}PO_4^{3-}$)	mg/L	
Dissolved gases	Carbon dioxide (CO_2)	mg/L	Scale formation, electrochemical corrosion, growth of aerobic autotrophic microorganisms
	Hydrogen Sulfide (H_2S)	mg/L	Scale formation, microbiologically mediated corrosion.
	Dissolved Oxygen (OD) (O_2)	mg/L	Scale formation, electrochemical corrosion, growth of aerobic autotrophic and aerobic heterotrophic microorganisms

Table 1.
Possible physicochemical parameters for the characterization of formation waters reuse or disposal. [Created by authors].

problems associated with water injection for reuse and EOR operations. This knowledge is also considered a starting point in preventing the risks of contamination of aquifers and soils, induced seismicity, and corrosion at the conduction lines. For a more complete characterization, it may be included a measurement of components used for the formulation of the hydraulic stimulation fluid that can cause problems in the pipelines.

Compatibility tests must be performed between the formation and the reinjection waters. These tests allow preventing future problems such as scale formations. Chemical speciation using these waters can be an initial step to determine possible affectations.

When production waters do not comply with the established maximum permissible limits of contaminants for discharge or reuse, according to local environmental regulations or operational parameters, it is necessary to select a set of processes to allow the removal of compounds to reach the water quality requirements. The removal levels will depend on the alternative use or reuse of the treated production waters. For example, for reuse in enhanced oil recovery or hydraulic fracturing, treatment requires low levels of removal of some contaminants at a relatively low cost, whereas, for reuse in crop irrigation, treatment requires high levels of removal of many contaminants, which have a high cost (Up to 7 times).

The expected results of produced water treatment processes in terms of the removal of specific contaminants vary depending on the process used. **Table 2** compiles quantitative information on the application intervals and expected removal percentages for fats and oils, suspended solids, and dissolved solids through different treatment processes.

The discharge or reuse standards, together with the characteristics of the production waters, define the processes that must be implemented to structure a treatment technology. In most applications, the main treatment needs will include one or more of the following levels: 1) remove greases and oils, 2) remove dissolved solids, 3) decrease BTEX concentrations, 4) decrease the biological oxygen demand of soluble organic compounds, 5) control elevated levels of volatile organic acids, 6) control suspended solids, 7) reduce brine volumes that require disposal, 8) control total and fecal

	Process	Application interval (mg/L)	Removal (%)
Greases & oil	Separators		
	API	500–20.000	90
	Corrugated Plate Interceptor	500–10.000	90
	Induced Gas Flotation (IGF)	500–1.000	96
	Hydrocyclone	300–500	95–99
	Centrifuge	100–10.000	93
	Filters		
	Porous media (walnut shell)	50–100	98
	Microfiltration (MF)	50–180	99
	Thermal (distillation)		
	Multi-effect	< 20	99
	Vapor compression	< 20	98
	Alternatives		
	Evaporation-crystallization	< 20	99
Total Suspended Solids (TSS)	Separators		
	API	< 1.000	50–75
	Corrugated Plate Interceptor	< 400	80
	Induced Gas Flotation (IGF)	< 200	85
	Hydrocyclone	< 200	65–80
	Centrifuge	30–300	65–80
	Filters		
	Porous media (walnut shell)	< 30	99
	Microfiltration (MF)	< 20	98
	Thermal (distillation)		
	Multi effect	< 10	100
	Vapor compression	< 10	100
	Alternatives		
	Evaporation-crystallization	< 10	100
Total Dissolved Solids (TDS)	Membranes		
	Reverse osmosis (RO)	1.000–45.000	> 99
	Electrodialysis-Reverse electrodialysis	500–40.000	99.5
	Thermal (distillation)		
	Multistage flash	5.000–50.000	> 99.9
	Multi-effect	1.500–100.000	> 99.9
	Vapor compression	1.500–200.000	> 99.9
	Membranes	500–250.000	> 99.5
	Freeze–thaw	> 5.000	> 94
	Alternatives		
	Ion exchange	< 750	95
	Capacitive deionization	500–5.000	99
	Evaporation-crystallization	300.000	> 99.9

Table 2.
Application intervals and removal percentages for some parameters through different production waters treatment processes. [Created by authors].

coliforms, 9) remove constituents of special interest (e.g. boron, which restricts the final use in, for example, irrigation), and 10) adjust the rate of sodium adsorption to avoid its retention (only for irrigation or shallow aquifer recharge). In recent decades, technologies have been developed for different levels of treatment of production waters that allow discharge or reuse standards to be achieved.

Finally, in operations that handle the extraction, reinjection, treatment, and recirculation of large volumes of water through porous media of high pressure and, in some cases, high temperatures, induced seismicity and the appearance of microseismic nests are inevitable.

In these cases, it is essential to manage events through holistic knowledge of the intervened porous medium and the main parameters of the technical operation, requiring at least 1) defining and accurately mapping the proximity of the crystalline basement, the geomechanical properties, and faults in the formations subject to injection; 2) design and adjust, when necessary, the well geometric arrangement and disposal rate of fluids, and 3) monitor the accumulated volume of such fluids.

Author details

John Fernando Escobar[1], Darío Naranjo-Fernández[1*], Sergio Lopera[2], Orlando Mauricio Quiroz[3], Alonso Ocampo[4] and German Zarate[5]

1 University of Antioquia, Medellín, Colombia

2 National University of Colombia, Medellín, Colombia

3 National University of Mar del Plata, Mar del Plata, Argentina

4 Gastim Technologies, Bogotá, Colombia

5 National Hydrocarbons Agency, Bogotá, Colombia

*Address all correspondence to: dario.naranjo@udea.edu.co

References

[1] Moncur M, Foley S, et al. Groundwater development for the oil industry. Journal of Canadian Petroleum Technology. 1999;**38**(11)

[2] Neff J, Lee K, DeBlois EM. Chapter 1. Produced water: Overview of composition, fates, and effects. In: Lee K, Neff J, editors. Produced Water: Environmental Risks and Advances in Mitigation Technologies. USA: Springer Science; 2011. pp. 3-54. DOI: 10.1007/978-1-4614-0046-2

[3] Thakur G, Satter A. Introduction. Integrated Waterflood Asset Management. U.S.A: PennWell Corporation; 1998

[4] Hadi S, Wahab MA, Nasron BM, Sharkawi WA, Latiff NA. Molecule to molecule M2M water injection performance review to achieve water injection excellence in PETRONAS. In: SPE Paper 197363; presented at the Abu Dhabi International Petroleum Exhibition & Conference, 11-14 November 2019. Abu Dhabi, UAE; 2019

[5] Castro R, Maya G, Mantilla J, Diaz V, Amaya R, Lobo A, et al. Waterflooding in Colombia: Past, present, and future. In: SPE Paper 169459; Presented at the SPE Latin American and Caribbean Petroleum Engineering Conference Held in Maracaibo, 21-23 May, 2014. Venezuela; 2014

[6] ANH. Production Statistics. 2022. Available from: https://www.anh.gov.co/es/operaciones-y-regal%C3%ADas/sistemas-integrados-operaciones

[7] ECOPETROL. Eficiencia operativa en el manejo del agua. 2022. Available from: https://www.ecopetrol.com.co/wps/portal/Home/sostecnibilidad/ambiental/gestion-integral-del-agua/

[8] Ministerio de Minas y Energía de Colombia. Centro de Transparencia. 2022. Available from: https://www.centrodetransparenciappii.org/YNC

[9] ANLA. Resolucion 0855. 2022. Available from: https://www.anla.gov.co/eureka/normatividad

[10] Collins AG, Wright CC. Chapter 6: Enhanced oil recovery injection waters. In: Chilingarian GV, editor. Developments in Petroleum Science. The Netherlands: Elsevier Science Publishers; 1985. pp. 151-221. DOI: 10.1016/S0376-7361(08)70568-5

[11] Yang M. Chapter 2. Measurement of oil in produced water. In: Lee K, Neff J, editors. Produced Water: Environmental Risks and Advances in Mitigation Technologies. USA: Springer Science; 2011. pp. 57-88. DOI: 10.1007/978-1-4614-0046-2

[12] Al-Ghouti MA, Al-Kaabi MA, Ashfaq MY, Da'na, D.A. Produced water characteristics, treatment and reuse: A review. Journal of Water Process Engineering. 2019;**28**:222-239. DOI: 10.1016/j.jwpe.2019.02.001

[13] Siagian U, Widodo S, Khoiruddin K, Wardani A, Wenten IG. Oilfield produced water reuse and reinjection with membrane. MATEC Web of Conferences. 2018;**156**:08005. DOI: 10.1051/matecconf/201815608005

[14] Guerra K, Dahm K, Dundorf S. Oil and gas produced water management and beneficial use in the western United States. In: Science and Technology Program Report No. 157. Bureau of Reclamation, USA: U.S. Department of the Interior; 2011. p. 113. Available from: https://www.usbr.gov/research/dwpr/reportpdfs/report157.pdf

[15] Veil J, Puder M, Elcock D, Redweik RJ. A White Paper Describing Produced Water from Production of Crude Oil, Natural Gas, and Coal Bed Methane. USA: U.S. Department of Energy; 2004. p. 87. Available from: https://publications.anl.gov/anlpubs/2004/02/49109.pdf

[16] Environmental Protection Agency. Assessment of the Potential Impacts of Hydraulic Fracturing for Oil and Gas on Drinking Water Resources, External Review Draft EPA/600/R-15/047a, [Internet]. 2015. Available from: https://cfpub.epa.gov/ncea/hfstudy/recordisplay.cfm?deid=244651

[17] Fetter CW, Boving T, Kreamer D. Contaminant Hydrogeology. Third ed. USA: Waveland Press; 2018. p. 647

[18] Veil JA. Chapter 29. Produced water management options and technologies. In: Lee K, Neff J, editors. Produced Water: Environmental Risks and Advances in Mitigation Technologies. USA: Springer Science; 2011. pp. 537-571. DOI: 10.1007/978-1-4614-0046-2

[19] Coonrod CL, Yin YB, Hanna T, Atkinson A, Alvarez PJJ, Tekavec TN, et al. Fit-for-purpose treatment goals for produced waters in shale oil and gas fields. Water Research. 2020;**173**:115467

[20] Esmaeilirad N, Terry C, Kennedy H, Prior A, Carlson K. Recycling fracturing flowback water for use in hydraulic fracturing: Influence of organic matter on stability of carboxyl-methyl-cellulose based fracturing fluids. Society of Petroleum Engineers Journal. 2016;**21**(4):1358-1369. DOI: 10.2118/179723-PA

[21] Torres L, Yadav OP, Khan E. A review on risk assessment techniques for hydraulic fracturing water and produced water management implemented in onshore unconventional oil and gas production. Science of the Total Environment. 2016;**539**:478-493. DOI: 10.1016/j.scitotenv.2015.09.030

[22] Boschee P. Produced and flowback water recycling and reuse: Economics, limitations, and technology. Oil and Gas Facilities. 2014;**3**(1):16-22. Available from: http://www.spe.org/ogf/print/subscribers/2014/02/07_Feat_Unconventional.pdf

[23] Hammer R, VanBriesen J. In fracking's Wake: New Rules Are Needed to Protect our Health and Environment from Contaminated Wastewater. USA: Natural Resources Defense Council (NRDC); 2012. p. 129. Available from: https://www.nrdc.org/sites/default/files/Fracking-Wastewater-FullReport.pdf

[24] Dake LP. The Practice of Reservoir Engineering. In: Series: Developments in Petroleum Science. 1st ed. Vol. 36. The Netherlands: Elsevier Science Publishers; 2001. p. 572

[25] Osorio JG, Peñuela G, Otalora O. Correlation Between Microseismicity and Reservoir Dynamics in a Tectonically Active Area of Colombia2008. Available from: https://www.researchgate.net/publication/254528732_Correlation_Between_Microseismicity_and_Reservoir_Dynamics_in_a_Tectonically_Active_Area_of_Colombia. DOI: 10.2118/115715-MS

[26] Healy J, Rubey W, Griggs D, Raleigh C. The Denver earthquakes. Science. 1968;**161**:1301-1310

[27] Raleigh C, Healy J, Bredehoeft J. An experiment in earthquake control at Rangely, Colorado. Science. 1976;**191**:1230-1237

[28] Brown MRM, Ge S. Distinguishing fluid flow path from pore pressure

diffusion for induced seismicity. Bulletin of the Seismological Society of America. 2018;**108**(6):3684-3686. DOI: 10.1785/0120180149

[29] Zang A et al. How to reduce fluid injection induced seismicity. Rock Mechanism Rock Engineering. 2018;**54**:475-493. DOI: 10.1007/s00603-018-1467-4

[30] Sharma MM, Yen TF, Chilingarian GV, Donaldson EC. Chapter 7. Some chemical and physical problems in enhanced oil recovery operations. In: Chilingarian GV, editor. Developments in Petroleum Science. The Netherlands: Elsevier Science Publishers; 1985. pp. 223-249. DOI: 10.1016/S0376-7361(08)70568-5

[31] Hibbeler J, Garcia T, Chavez N. An integrated Long-Term Solution for Migratory Fines Damage. In: Paper SPE 81017 Presented at the 2003 SPE Latin American and Caribbean Petroleum Engineering Conference. Trinidad; 2003

[32] Civan F. Chapter 8, Particulate Process in Porous Media Reservoir Formation Damage. Houston: Gulf Publishing Co.; 2000. pp. 140-163